Molecular Biology Intelligence Unit

Recombinational Repair of DNA Damage

MOLECULAR BIOLOGY INTELLIGENCE UNIT

RECOMBINATIONAL REPAIR OF DNA DAMAGE

Andrei Kuzminov

Institute of Molecular Biology
University of Oregon
Eugene, Oregon

CHAPMAN & HALL
I(T)P An International Thomson Publishing Company

New York • Albany • Bonn • Boston • Cincinnati • Detroit • London • Madrid • Melbourne • Mexico City • Pacific Grove • Paris • San Francisco • Singapore • Tokyo • Toronto • Washington

R.G. LANDES COMPANY
AUSTIN

Molecular Biology Intelligence Unit

RECOMBINATIONAL REPAIR OF DNA DAMAGE

R.G. LANDES COMPANY
Austin, Texas, U.S.A.

Printed in the U.S.A.

Please address all inquiries to the Publishers:
R.G. Landes Company, 909 Pine Street, Georgetown, Texas, U.S.A. 78626
Phone: 512/ 863 7762; FAX: 512/ 863 0081

North American distributor:
Chapman & Hall, 115 Fifth Avenue, New York, New York, U.S.A. 10003

CHAPMAN & HALL

U.S. and Canada ISBN: 0-412-10671-X

Library of Congress Cataloging-in-Publication Data

CIP data applied for but not received by time of publication.

Publisher's Note

R.G. Landes Company publishes six book series: *Medical Intelligence Unit, Molecular Biology Intelligence Unit, Neuroscience Intelligence Unit, Tissue Engineering Intelligence Unit, Environmental Intelligence Unit* and *Biotechnology Intelligence Unit.* The authors of our books are acknowledged leaders in their fields and the topics are unique. Almost without exception, no other similar books exist on these topics.

Our goal is to publish books in important and rapidly changing areas of bioscience for sophisticated researchers and clinicians. To achieve this goal, we have accelerated our publishing program to conform to the fast pace in which information grows in bioscience. Most of our books are published within 90 to 120 days of receipt of the manuscript. We would like to thank our readers for their continuing interest and welcome any comments or suggestions they may have for future books.

Deborah Muir Molsberry
Publications Director
R.G. Landes Company

Dedication

To my mother, whose dream became my destiny.

CONTENTS

PREFACE

This book is about the mechanisms cells employ to repair potentially lethal DNA damage. These mechanisms are the same as the mechanisms of homologous recombination, justifying the term "recombinational repair". A book that would fill the conceptual gap between the historically separated fields of Genetic Recombination and DNA Repair is overdue. Genetic recombination is widely perceived as a mechanism by which organisms increase their fitness on the evolutionary scale, while DNA repair is treated as mundane transactions of DNA maintenance. This book attempts a synthesis of these two fields by analyzing possible requirements of complex repair reactions and matching the known recombinational enzymes into the recombinational repair pathways.

Recombinational repair is a processes of such complexity that it took three quarters of a century of collecting observations and decades of experimenting to conceive its internal logic. In contrast to its simpler cousin, one-strand DNA repair, whose role in mending the bulk of DNA lesions was apparent from the very beginning, recombinational repair was for a long time studied innocently through one of its infrequent consequences, homologous recombination. The idea that cells employ the mechanisms of homologous recombination to repair complex DNA lesions is not new, but there is a shortage of efforts to fit the ever accumulating knowledge about recombination into the requirements of repair. Nevertheless, the great puzzle of recombinational repair is no longer a pile of "individual pieces" — some parts are already half-assembled. Efforts to connect these separate half-assembled parts are hampered by the fact that enzymes of recombinational repair frequently exhibit overlapping activities. Consequently, deducing the mode of in vivo action of these proteins only on the basis of their in vitro activities could be perilous.

This book is an example of a different approach. The experimentally established principles were used to predict the likely requirements of a particular step in a recombinational repair reaction; these requirements then served as the guidelines for selecting candidates that could catalyze the step. In effect, an outline of the final picture was first visualized and then filled in with available pieces. The unavoidable consequence of this approach was the omission of some of the facts of the field. I can only hope that these omissions were justified by the heuristic value of the resulting picture.

The book classifies DNA lesions, traces the evolution of recombinational repair models, introduces the concept of synapsis as the essence of recombinational repair, discusses the reactions catalyzed by synaptases and possible mechanisms of homology search, and offers a perspective on the SOS response. On this basis, the major prokaryotic pathways of recombinational repair are analyzed and, lastly, the prokaryotic paradigm is compared with what is known about recombinational repair in eukaryotes. The book will be helpful both to newcomers to the field (by providing a systematized view) and to old-timers (by giving a fresh look to old facts). For those who do not have time even to eat their breakfast, but still want to get the punch line, it will be sufficient to browse through the highlights and to study the illustrations. To help readers to travel freely around the subject, cross-references (for example (7.1), meaning "chapter 7, part 1") are provided.

Study of homologous recombination was instrumental in developing the concept of recombinational repair, and it is unfortunate that the format of the book and time constraints have not allowed me to discuss this exciting topic in any detail. A serious coverage of homologous recombination would have taken another couple of books and would have constituted an unnecessary repetition of the existing excellent books and reviews on that subject (see "Selected Reading").

Acknowledgment

The book would be impossible without Frank Stahl, who was royally generous in everything he did, from revealing my misconceptions and misrepresentations of the facts to supporting me financially during the writing. Both he and Lynn Thomason helped with editing the text.

CHAPTER 1

INTRODUCTION

Every cell has a genome, a self-replicating organizer of cellular metabolism. The genome stores and manipulates vital information employing two forms of nucleic acids — RNA and DNA. These long polymers of nucleotides encode the entire variety of cellular reactions, including their own reduplication.

The genome fulfills two major functions: (1) storage and duplication of information; (2) regulation of expression and transfer of information. In the first living creatures, both functions as well as biocatalysis itself were probably carried out by various RNA molecules (for sketches see refs. 1, 2). While the chemical nature of RNA makes it universally useful in both biocatalysis and information transfer, it also renders RNA rather unstable — the bonds between ribonucleotides are prone to hydrolysis. Therefore, RNA was certainly not the best choice for the long-term informational storage.

As genomes grew in size, the problem of stability of the informational molecule must have turned into a roadblock precluding further progress. The solution was found in substituting double-stranded DNA for RNA as the main information storage molecule because of its superiority to RNA in chemical stability. In modern organisms, RNA retains its role in information transfer (messenger RNA, transfer RNA) and catalysis (ribosomal RNA, small nuclear RNA). The susceptibility of RNA to hydrolysis became a useful feature, allowing sensitive regulation of the cellular enzymatic milieu via mRNA degradation.

Although DNA is relatively stable compared to RNA, it, like any macromolecule, is subject to constant chemical modification and breakage. Repair of these lesions is essential for DNA to carry out its roles as template for transcription and its own reduplication. In the course of evolution, a complex enzymatic machinery has evolved to maintain this centrally important molecule in usable form.

1.1. TWO-STRAND DNA REPAIR: A HELP FROM SISTER, RECOMBINATION AND MODELS

As the DNA lesions are of various nature and complexity, so the enzymology of their repair is also diverse. If a DNA lesion affects only one strand of a duplex DNA, it requires one type of repair reactions. If a lesion strikes both DNA strands across each other, it needs more sophisticated repair. The initial methods of studying these two types of repair were fundamentally different, resulting in separation of the two fields for decades.

1.1.1. Damage-Reversal, One-Strand Repair and Two-Strand Repair

Repair of some DNA modifications simply reverses the damage, returning DNA directly to its original state. For instance, photolyase, using near UV-visible light, splits UV-induced pyrimidine dimers (reviewed in ref. 3). Another example is the suicidal Ada protein of *E. coli*, which transfers a methyl group from a modified base O^6-methylguanine to itself (reviewed in ref. 4). However, repair of a lesion frequently leads to loss of a portion of the damaged DNA strand. Double-strandedness of DNA, although at first glance a redundancy, provides the means for repairing such a gap. The damaged segment in one strand can be resynthesized using the intact complementary strand as a template.

One example of such a strategy is in the repair of modified bases that do not cause DNA distortion. The so-called "base-excision repair" acts with an unexpected precision — an enzyme called DNA glycosylase removes a modified base to produce an abasic site, the phosphodiester bond is broken to the 5'-side of the site, and repair is completed by a one base long nick translation by DNA polymerase.[5] Another repair system, the nucleotide excision repair, deals with DNA lesions that do cause distortion of the double helix. An excinuclease removes a segment of a single strand (12-13 nucleotides in *E. coli*, 25-26 nucleotides in humans) that has such a lesion in the middle; the removed strand is replaced by repair synthesis (reviewed in ref. 6). The third repair system, methyl-directed mismatch repair, can liberate up to a thousand nucleotides from one strand in its efforts to remove a single mismatch arising during DNA replication (reviewed in ref. 7). We will call a damage to one DNA strand a one-strand lesion, and repair of such a damage a one-strand repair.

The double-stranded nature of DNA allows the repair of the major fraction of DNA damage using the complementary strand as a template. Sometimes, however, both DNA strands are damaged opposite each other. Typically, such a situation develops when a replication fork encounters an unrepaired lesion in a template DNA strand. Another situation is a disruption of both strands in the same position which introduces a "double-strand break" in DNA. With such a two-strand damage,[8] the doublestrandedness of DNA is of little help. To repair two-strand lesions without the loss of sequence information, a cell needs a higher level of redundancy, an extra homologous sequence whose strands could be used to fix both DNA strands of the damaged sequence. The principle of two-strand repair is depicted in Figure 1.1.

An affected region is paired (synapsed) with an intact homologous region using complementarity of the corresponding DNA strands (Fig. 1.1B). The generated joint molecule is "resolved" by precisely placed single-strand cuts yielding two new DNA molecules each one containing a one-strand lesion (Fig. 1.1C). Now, the damaged strands can be mended by one-strand repair using the complementary strands as templates (Fig. 1.1D). Thus, the strategy of two-strand repair is to convert a two-strand lesion into a pair of one-strand lesions by synapsis with an intact homologous DNA sequence. This point is illustrated by the long established requirements for duplicate genomes to repair double-strand breaks in both bacterium *Escherichia coli*[9] and in yeast *Saccharomyces cerevisiae.*[10,11]

1.1.2. Genetic Analysis and Biochemical Analysis

A contemporary biologist deciphers a complex process, like a metabolic pathway, with two equally powerful and complementary means. One is genetic analysis, which starts with devising a scheme for the isolation of mutants with a particular phenotype. After the discovered mutants are mapped to distinct loci (genes), one can do epistatic analysis, constructing various double mutants and testing them for the phenotype of interest.[12] If a phenotype of a double mutant is not different from the phenotype of parental single mutants, the conclusion is that both genes control the same metabolic pathway. If a phenotype of a double mutant shows an additive effect, this is interpreted to signal the existence of two nonoverlapping and noncompeting pathways. At last, if a phenotype of a double mutant displays a synergistic effect, it is concluded that the genes control distinct but competing pathways. The last, most interesting stage of genetic analysis is determining at what stage the process of interest is blocked in a particular mutant; such an analysis often requires help from other methods.

The complementary way of studying a metabolic pathway is biochemical analysis. It starts with devising an assay for a particular activity and looking for this activity in crude extracts of the cells of choice. Eventually, one can isolate and characterize an enzyme possessing the sought activity. Repeated many times with different steps of a metabolic pathway, biochemical analysis can reveal all the activities of the pathway. It culminates in the demonstration that a combination of the individual activities in one test tube catalyzes the entire pathway, from the initial substrates to the final products.

1.1.3. Biochemistry versus Genetics: One-Strand Repair versus Recombinational Repair

The mechanisms for the repair of one-strand lesions are not easily examined using genetic analysis, but succumb better to biochemical analysis. Genetic analysis has no problem identifying a few genes controlling a particular one-strand DNA repair reaction, but falls short

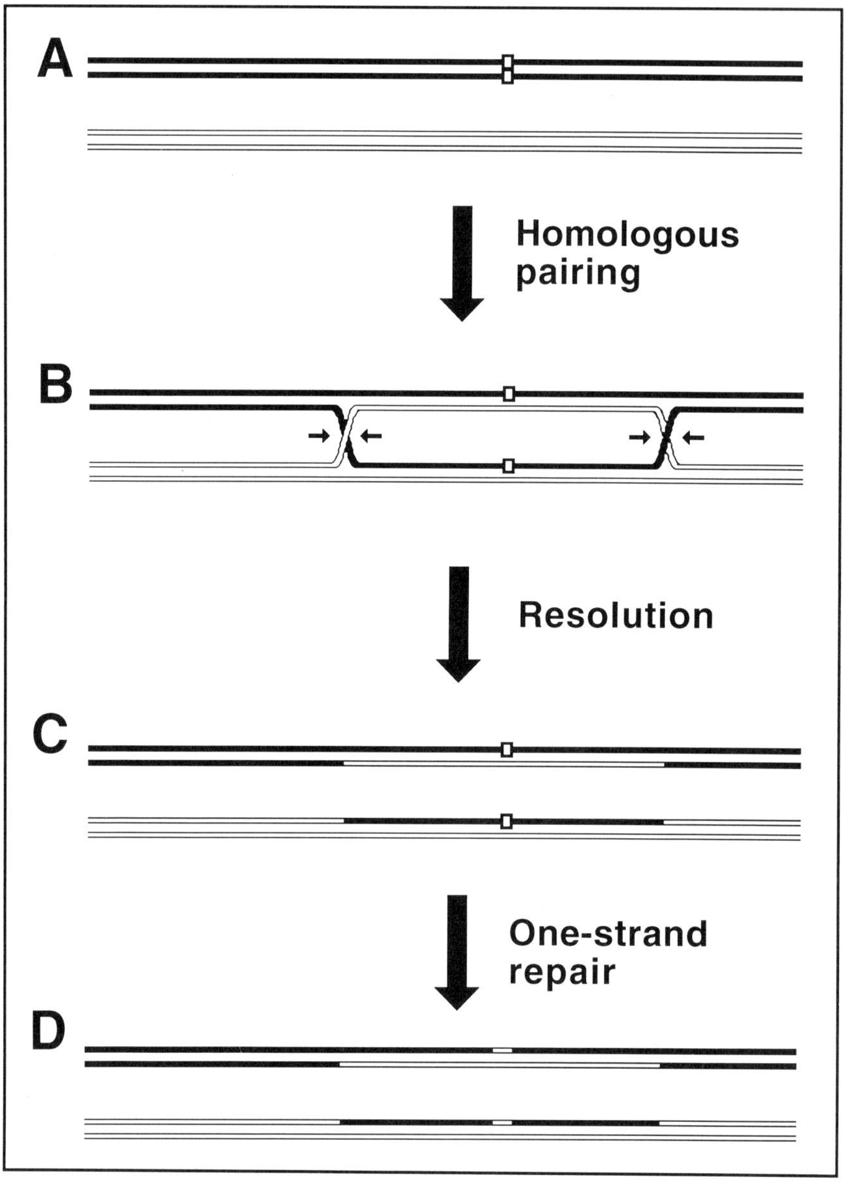

Fig. 1.1. The idea of two-strand repair. DNA lesions are drawn as small open rectangles. (A) A DNA sequence which suffered two-strand damage is shown side-by-side with a homologous intact sequence. (B) Two sequences have exchanged strands in a homologous region, converting the two-strand lesion into a pair of one-strand lesions. (C) Resolution of the junctions (direction of the resolution is shown in "B") frees chromosomes from each other. (D) Excision repair removes one-strand lesions, completing the overall repair reaction. Note that if black and white "parental" DNA are not identical, the resulting chromosomes may become "recombinant".

of arranging them in a metabolic pathway. One-strand repair reactions usually comprise a few poorly discernible steps, and often are completed in a single stroke, making their step-by-step characterization by genetic analysis impossible. On the other hand, biochemical analysis of one-strand repair is relatively easy. The biochemistry of one-strand repair is not perplexing, the key steps being catalyzed by single, though sometimes multisubunit, enzymes. The cells usually have a considerable capacity for this type of repair, which facilitates isolation of individual activities. If a particular DNA modification is repaired in crude extracts, one has an immediate assay for the isolation of individual components for such a repair. After purification of the individual components, it is doable (still requiring up to a decade of work) to combine all of them in a test tube with a modified DNA and to get out an intact DNA as a product.

In contrast to one-strand repair, the repair of two-strand lesions is too complicated to be deciphered solely by means of biochemical analysis. Two-strand DNA repair is an intricate multistage process in which individual components are often parts of a bigger enzymatic machine. Besides, cells have a limited capacity for this type of repair, and many of the participating enzymes are specifically maintained at very low concentration because they are poisonous for the cell. Fortunately, this is exactly the situation where genetic analysis feels at ease.

To study two-strand repair genetically, two principal assays are used. One is to follow the survival of cells treated with DNA-damaging agents (ionizing radiation, UV-light, Mitomycin C). Mutants in two-strand repair (as well as many mutants in one-strand repair) exhibit a decreased survival after treatment with such agents. The other genetic assay is to monitor the formation of new chromosomes resulting from alternative resolution of the joint molecules (Fig. 1.1B and C). Indeed, a joint molecule can be redrawn[13] to show that the junctions holding two DNA duplexes together are able to isomerize (Fig. 1.2A and B). This isomerization of the junctions permits two equally legitimate ways of resolving each junction (shown by numbers beside the arrows, Fig. 1.2B). If the resolution is random, in 50% of the cases the participating chromosomes will exchange shoulders, forming two "recombinant" chromosomes (Fig. 1.2C). Progeny carrying recombinant chromosomes can be scored genetically as having two traits that initially resided on separate ("parental") chromosomes.

Thus, the formation of recombinant chromosomes should be one of the consequences of two-strand DNA repair. The formation of recombinant chromosomes was first observed microscopically as chiasmata in pairs of homologous chromosomes during meiosis[14] and then, at the beginning of this century, was also demonstrated genetically and called "crossing-over".[15] Researchers started making use of crossing-over to map genes on the chromosomes long before DNA was shown to be the informational component of the chromosomes.[16] The

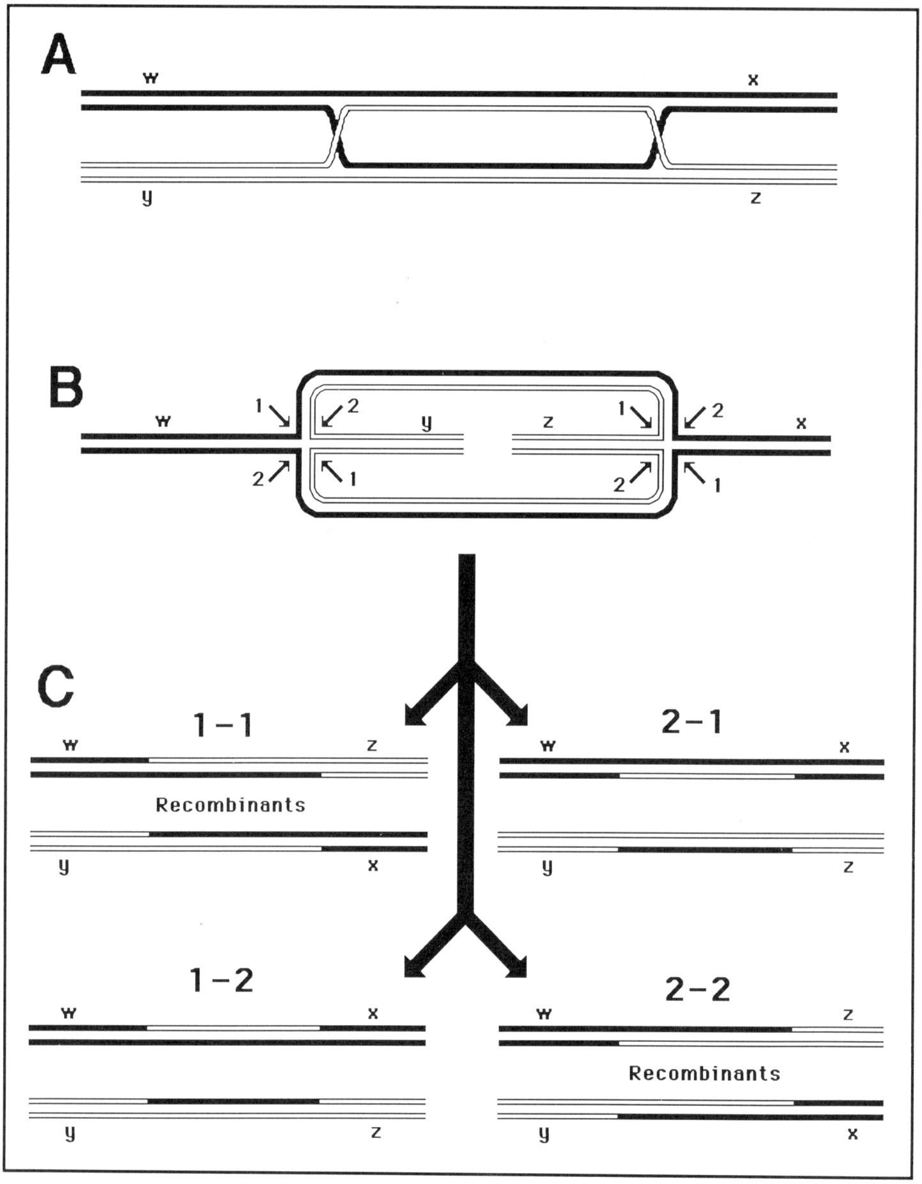

Fig. 1.2. Four ways joint molecules held by two junctions can be resolved. Small letters: w, x, y, z – designate unique sites which serve as markers on the homologous chromosomes. A junction is resolved by two symmetrical single-strand cuts (small black arrows in "B") across each other. Each such diagonal pair of cuts is numbered for each junction. If junctions freely isomerize and are resolved independently of each other, four outcomes of resolution are predicted. (A) A double junction as shown in Fig. 1.1B. (B) The same junction isomerized so as to show both junctions in open planar configuration. (C) The four resolution outcomes, numbered according to the resolution options realized at the left and the right junctions.

correlation of chiasmata and genetic cross-overs, which had been always suspected, was eventually confirmed.[17] It is believed now that recombinational interactions in meiosis are needed to ensure proper disjunction of homologous chromosomes.[18] However, prokaryotes lack meiosis yet are proficient in genetic recombination, suggesting that its primary benefit for them is different. When Harm in 1964 isolated a mutant of bacteriophage T4 defective in both genetic recombination and DNA repair, he hypothesized the existence of a repair pathway that has some mechanism in common with genetic recombination.[19] A discovery of similar mutants in *E. coli* followed shortly,[20,21] and by 1971 the concept of **recombinational repair** of DNA damage had been formulated by Howard-Flanders and colleagues.[22] Nevertheless, the idea that genetic recombination is a manifestation of two-strand DNA repair, although surfacing periodically, did not gain widespread support until fairly recently (reviewed in ref. 23).

Much of what we know about recombinational repair derives from studies aimed at understanding genetic recombination. We say that recombination happened when, as a result of sojourning of two DNA sequences together in one cell, there appears a hybrid sequence composed of the sequences from different parental chromosomes. Evolutionists long argued that genetic recombination allows a population to diversify and, consequently, to conquer new ecological niches and to increase its chances for survival during an environmental shift. However, evolutionary biologists originally thought of genetic recombination mostly as a random assortment of the parental chromosomes in the progeny, genetic crossing-over contributing only modestly to this indeed powerful mechanism for creating new parental gene combinations.[24] Molecular geneticists, on the other hand, considered homologous recombination strictly in terms of swapping segments between homologous chromosomes, but were slow to come up with alternative benefits for "their" recombination. Hence, for a long time homologous recombination (in the molecular meaning of this term), although associated with some risk for a single organism (in order to recombine, chromosomes must be broken), was thought to be selected for its ability to promote adaptive flexibility.[25] Yet, homologous recombination, although obviously important for proper chromosome segregation during meiosis in many organisms, is only an occasional consequence of recombinational repair, the process of extreme, immediate concern for an individual organism.[26] Recombinational repair usually mends a damaged chromosome with a help of an identical sister chromosome, and can be detected genetically only when a homologous (not identical) chromosome is used instead of a sister chromosome for repairing a two-strand DNA lesion. However, recombinational repair interactions between sister chromosomes are revealed by cytological methods as spectacular patterns of sister-chromatid exchange developing in response to DNA damage (7.2.2).

1.1.4. Milestones in the Evolution of Recombinational Repair Models

As the understanding of the molecular events underlying genetic recombination became clearer, models of homologous recombination converged upon those for recombinational repair. The former were designed mostly to explain reciprocal exchanges (crossing-overs) and associated nonreciprocal events (conversions) observed during meiosis in eukaryotes, while the latter were proposed to account for repair of two-strand DNA lesions. These two separate lines of development were finally united by the double-strand break repair model for meiotic recombination.

Homologous recombination has always fascinated geneticists. Early principles, inevitably very general, were formulated on the basis of microscopic observations and breeding experiments. In 1909, Janssens[27] proposed his "chiasmatype" model for crossing-over — a "break-and-join" swapping between regions of the homologous chromosomes (Fig. 1.3). A bold alternative, proposed by Belling in 1931[28,29] and also ascribed to Sturtevant[30] (and discussed by Lederberg[31]) was a "copy-choice" scheme (Fig. 1.3). Belling thought it is too much to ask homologous chromosomes to break simultaneously in exactly the same locations in preparation for exchange. In his scheme, a recombinant chromosome is copied partly from one parental chromosome and partly from the other, the switch happening in the region of shared homol-

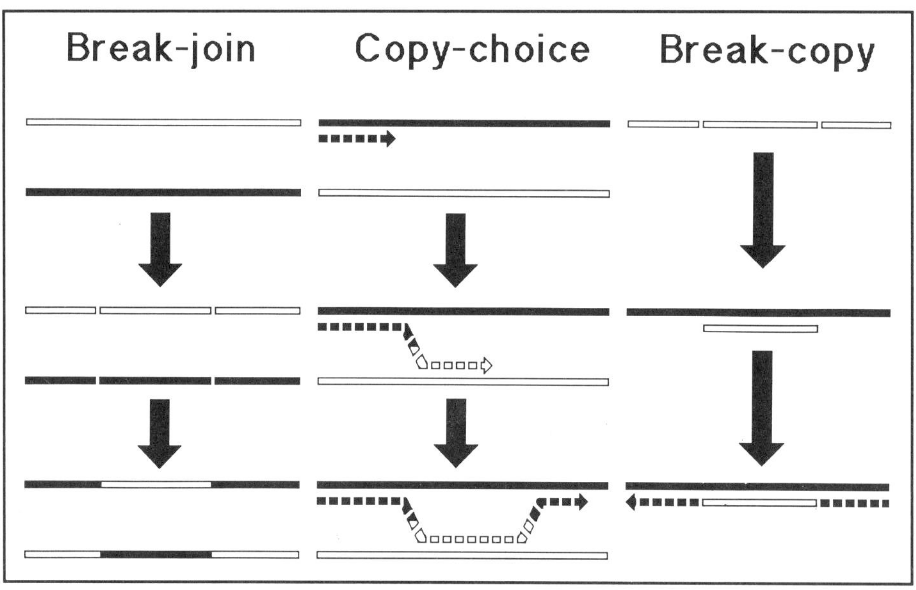

Fig. 1.3. First ideas about how recombinant chromosomes could be formed. Preexisting DNA duplexes are shown as either filled or open single lines; newly-synthesized duplexes are shown as dashed single lines. Explanations in the text.

ogy. A hybrid principle, which is now called "break-and-copy", envisioned first a break in one molecule and then copying the missing part from a homologous intact molecule.[31]

These early principles, though skeletal, provided a framework for thought. In the 1950s, two discoveries laid down new rules for designing recombination schemes. One was the 1953 Watson and Crick proposal for the molecular structure of duplex DNA;[32,33] the other was the 1958 demonstration by Meselson and Stahl of the semiconservative nature of DNA replication.[34] In 1961, models detailing break-join and break-copy recombination in bacteriophages were proposed by Meselson and Weigle[35] (Fig. 1.4). In contrast to many subsequent molecular models of recombination, these first schemes were not accompanied by a detailed verbal description. However, careful study of the figures reveals that at least three major elements of contemporary recombinational repair mechanisms were featured. First, Meselson and Weigle had recognized (preceded by Stahl[36] and Barricelli[37]) that, in order to recombine, DNA has to be damaged (broken). Second, in their break-copy scheme, the broken end invades the intact homologue to start DNA synthesis. And third, in the same break-copy scheme, we see a ubiquitous element of contemporary models — a four-way junction (regarding the resolution of which the authors were silent).

In 1963, Whitehouse, seeking an alternative to "copy-choice", proposed the first "molecular" break-join model of crossing-over.[38] He postulated that the initial breaks in the two participating homologous chromosomes were single-stranded, at noncoincidental positions and on chains of opposite polarity. The unraveling of the single-strands then allowed them to anneal with each other, securing interactions between homologous chromosomes (Fig. 1.5). This ingenious model has only a historic value nowadays, since none of its elements has been confirmed by experiment.

A model, suggested in 1964 by Holliday[39] fared better. In his model, Holliday proposed the contemporary view of the late stages of certain types of recombinational repair (Fig. 1.6). The centerpiece of the model was the same four-way junction that Meselson and Weigle had proposed without comment. The plausibility of such junctions was later supported with molecular model-building[40] while the junctions themselves were eventually detected in vivo in DNA molecules undergoing homologous interactions,[13] and were named "Holliday junctions" after their inventor.* Holliday thought that the initial breaks on the participating homologous chromosomes are concerted and even sequence-specific. For a variety of reasons, Holliday's model gave way to newer

* *Naming the four-way junctions after Holliday instead of Meselson and Weigle yet again demonstrates that "we tend to ascribe an idea not to a person who proposed it first, but to the one who first treated it seriously."*

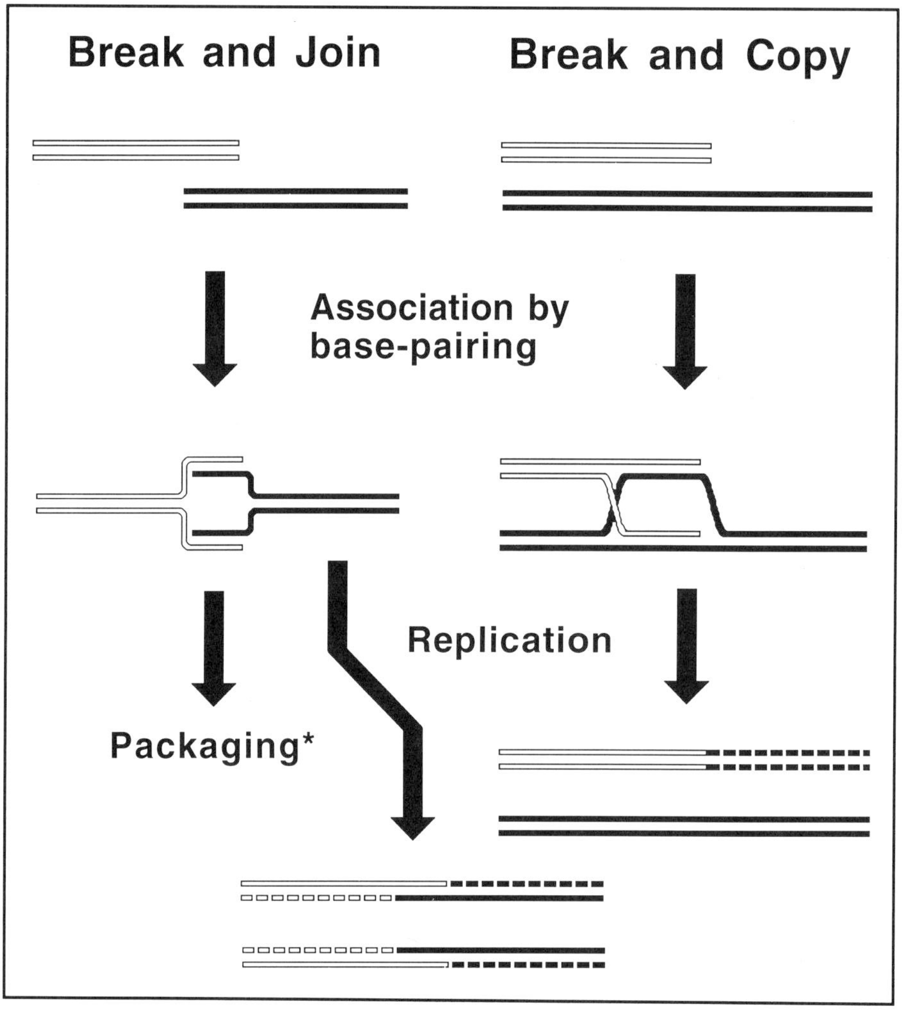

Fig. 1.4. The break-join and break-copy models by Meselson and Weigle (1961). DNA duplexes are shown by double lines. Explanations in the figure and in the text.
**It was assumed at that time that such a structure can be packaged directly, without the resolution of the overlapping strands and formation of the regular DNA duplex (Frank Stahl, personal communication).*

models in which initiating breaks occur on one chromosome, and the homologue is subsequently involved into the strand exchange.

1966 witnessed the appearance of two very successful models of genetic recombination, both of the break-join type, which have survived without major changes all these years. Fox[41] offered a model for bacterial transformation (recombination between a chromosome and an introduced DNA fragment (2.1.2)) in which he proposed: (1) strand invasion as a critical early stage of recombinational repair and (2) specific ways of resolving the resulting three-way structure (Fig. 1.7). Current

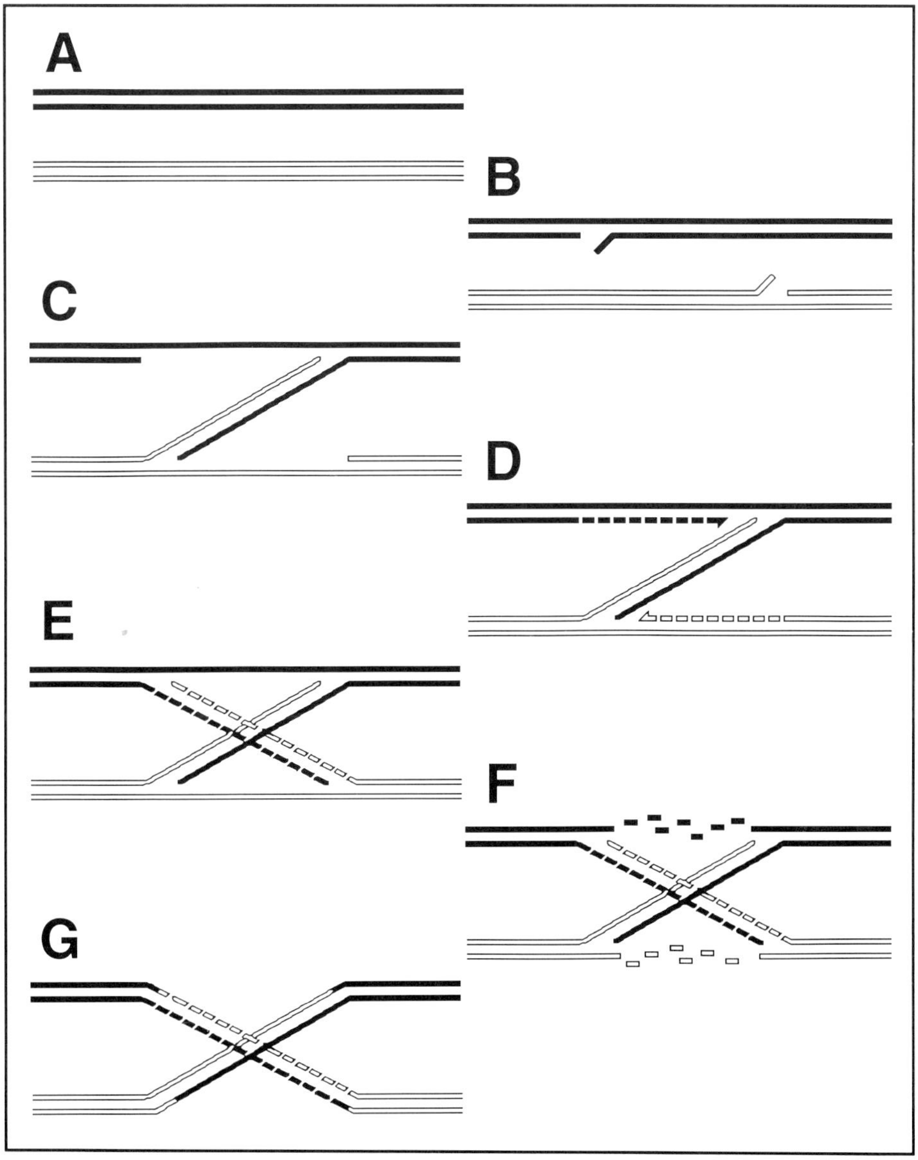

Fig. 1.5. The model of crossing-over by Whitehouse (1963). Preexisting DNA strands are shown as either filled or open lines; newly-synthesized strands are shown as dashed lines. (A) Two intact homologous DNA duplexes. (B) Single-strand breaks in complementary strands in different duplexes allow the unraveling of the strands. (C) The unraveled complementary strands from different duplexes anneal to form a bridge. (D) The single-stranded gaps are filled in by DNA synthesis. (E) The newly-synthesized strands are similarly unraveled and annealed, forming a cross. (F) The strands that do not participate in the cross are degraded. (G) The single-strand interruptions are repaired, completing the formation of a cross-over.

break-invade-join models (see, for example ref. 42), for which the Fox model is a prototype, adequately describe the type of recombination observed when DNA synthesis is artificially blocked. Looking back, it is odd to realize that it took another 10-15 years to couple the Holliday-64 and the Fox-66 schemes into the major current models of recombinational repair.

In the same year, 1966, Thomas proposed a distinctive model of break-join recombination which is another early scheme that is still

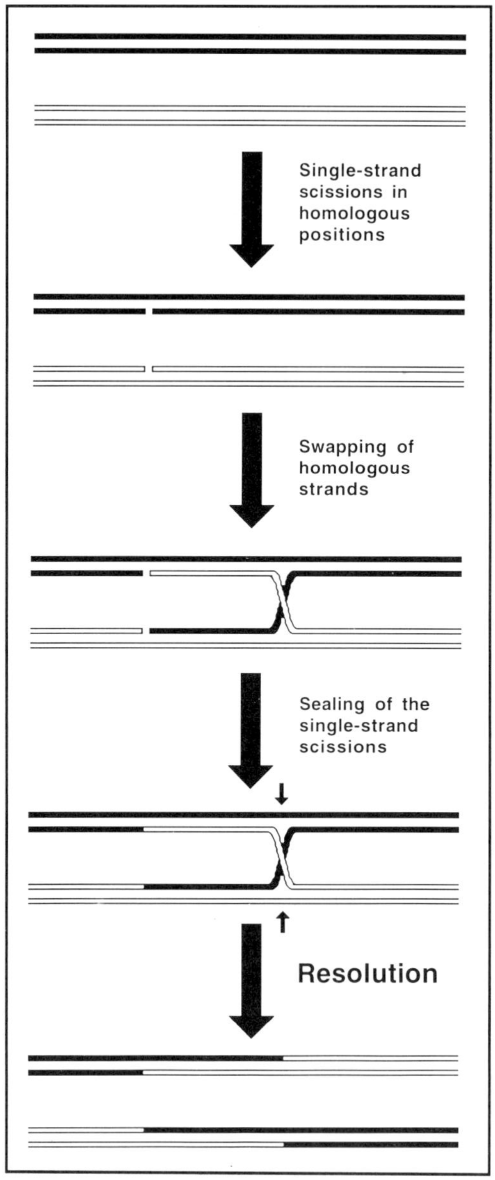

Fig. 1.6. The model of homologous recombination by Holliday (1964). The direction of Holliday junction resolution is shown with small arrows. For clarity, only one resolution product (cross-over) is shown. Explanations in the figure and in the text.

basically unchanged.[43] It envisions two simultaneous double-strand breaks in two molecules at different locations. The single-strand degradation of the same polarity (proposed later, see ref. 44) unravels complementary overlapping tails, which eventually anneal with each other (Fig. 1.8). Initially, this elegant model did not enjoy attention it deserved, may be because it seemed applicable only to recombination occurring among multiple genomes of an intracellular bacteriophage population[45] (although it was creatively used to account for recombinational repair of double-strand breaks in yeast chromosome[46]). However, today this mechanism is the darling of those who study repair of double-strand breaks in notoriously repetitive genomes of higher eukaryotes. It will be considered in detail in chapter 6.

In 1969, Boon and Zinder[47] proposed a model for recombination in bacteriophages, starting with a double-strand break in one molecule which is repaired with the use of an intact homologous molecule (Fig. 1.9). Invasion of one end of the double-strand break into the homologous chromosome establishes a replication fork which is broken after some progress. The elongated end can now recombine with

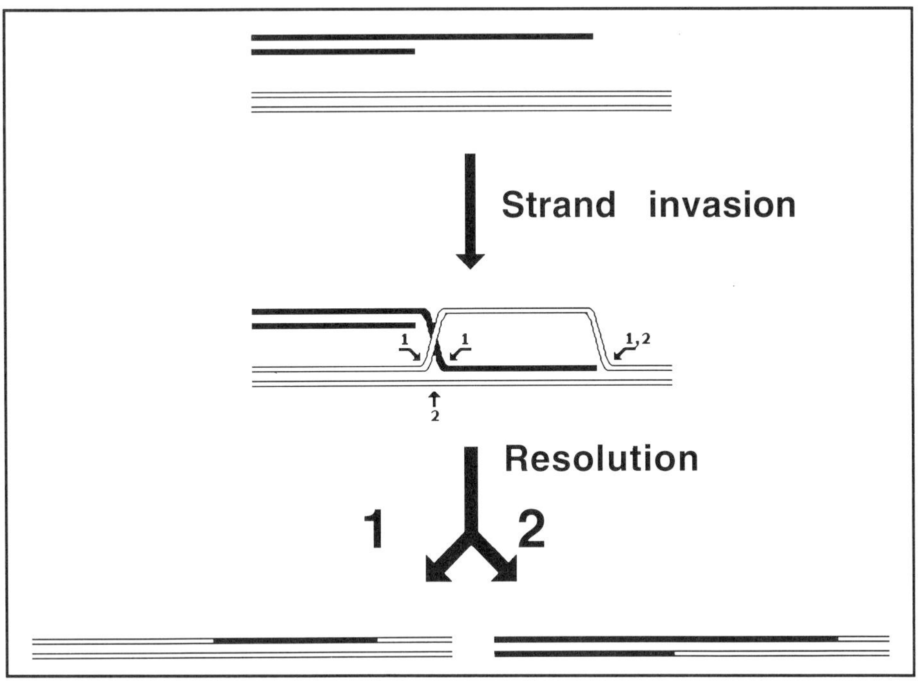

Fig. 1.7. The model of transformation by Fox (1966). The broken DNA molecule is shown in black, while the intact homologous duplex is shown in white. Single-strand scissions are shown by small numbered arrows. To receive the final product #1, one has to make all three incisions numbered "1" in the central panel. To get the final product #2, two correspondingly numbered incisions have to be introduced. The nicks and gaps are then repaired by DNA polymerase + DNA ligase.

the other end to repair the double-strand break. None of these elements in itself was original (invasion priming DNA replication was proposed by Meselson and Weigle (1961); invasion causing breakage was suggested by Fox (1966)), but the combination of them allowing complete double-strand break repair with the help of a homologous chromosome foreshadowed future trends.

In 1971 Hotchkiss,[48] possibly influenced by Paszewski,[49] suggested that meiotic recombination begins with a single-strand break in one of the homologues (Fig. 1.10). After the invasion of the 3'-end of the

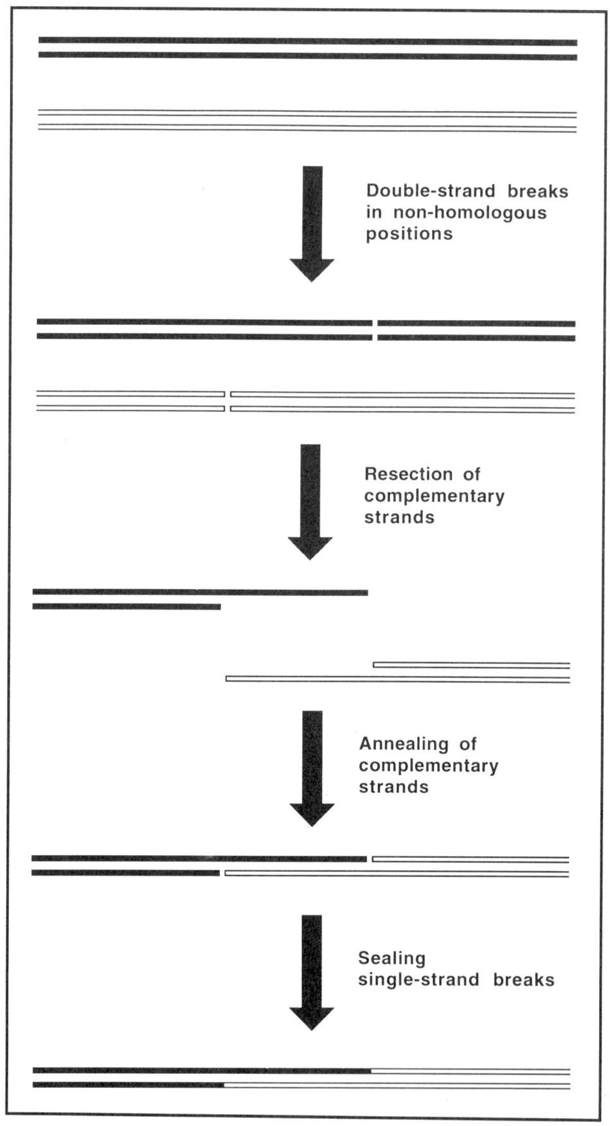

Fig. 1.8. The model of double-strand break repair by Thomas (1966-67). Explanations in the figure and in the text.

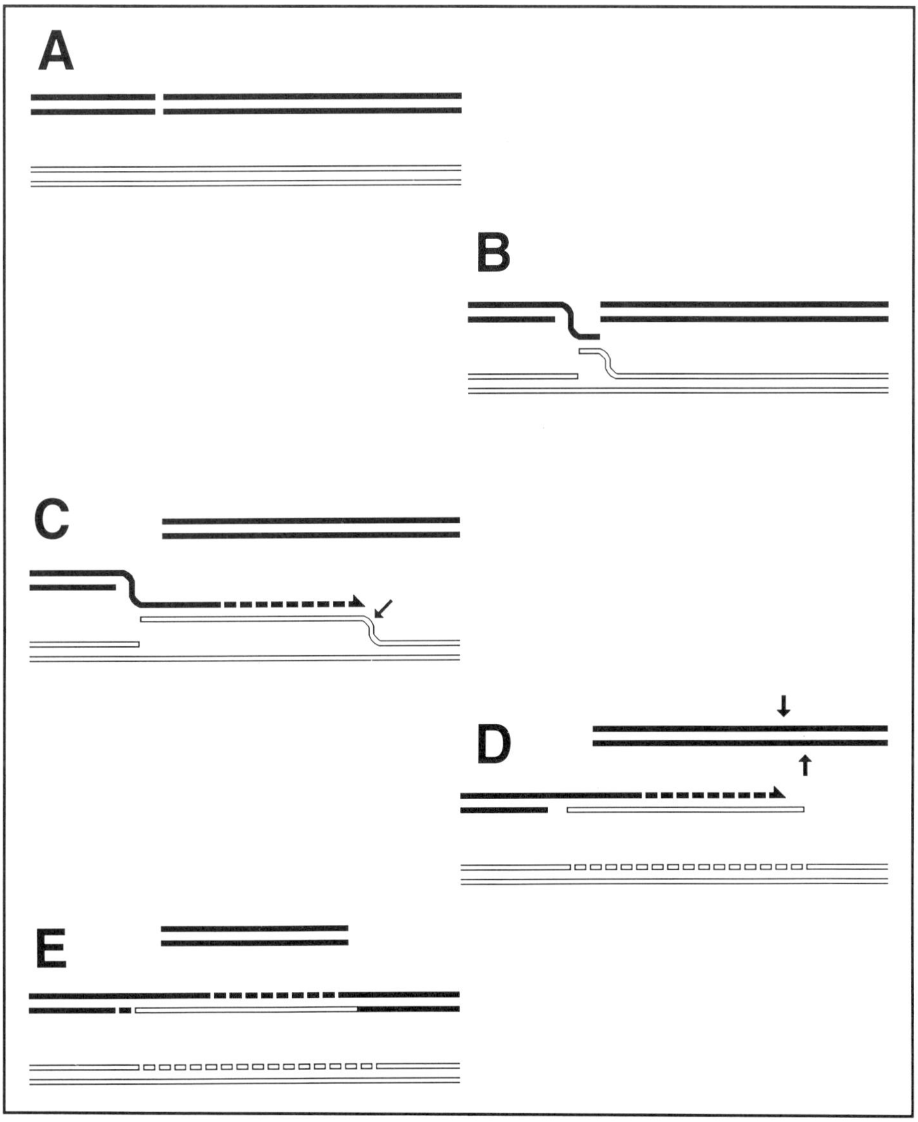

Fig. 1.9. The model of recombination via double-strand break repair by Boon and Zinder (1969). Newly-synthesized DNA is shown as interrupted lines. Required single-strand scissions are shown by small arrows. (A) Homologous duplexes, one broken and one intact. (B) Strand invasion accompanied by incision of the displacement loop – a replication fork is formed. (C) DNA synthesis elongates the end. (D) The elongated double-strand end is released as a result of breakage of the replication fork. (E) The double-strand end invades into the other end of the original double-strand break to reconstruct the duplex. A duplex linear fragment is released.

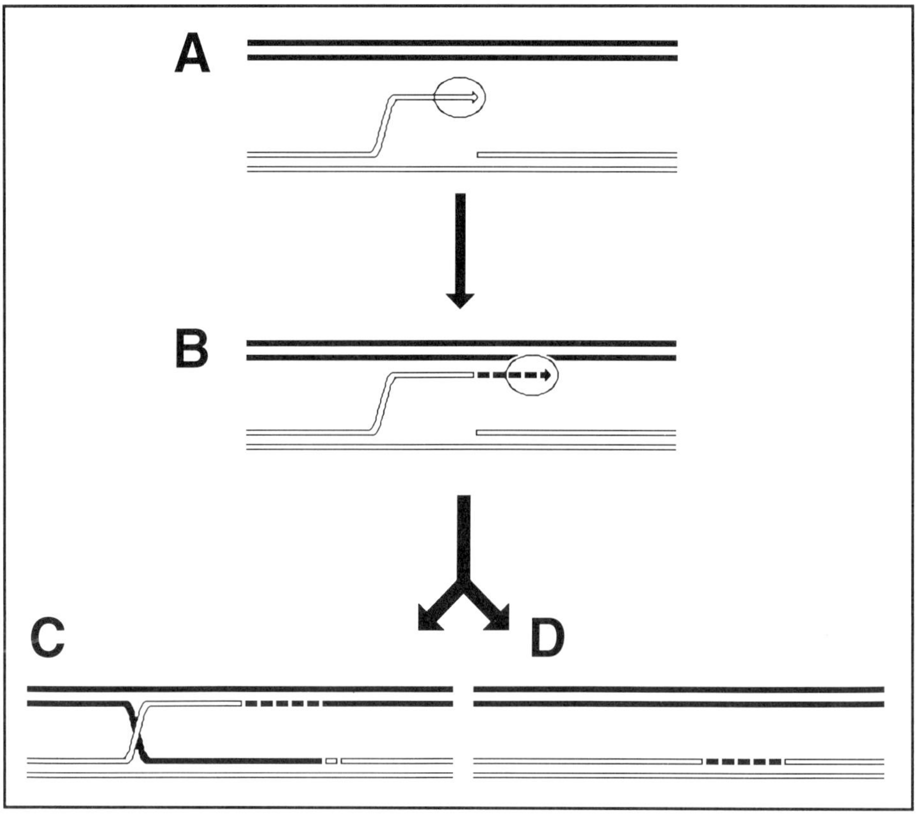

Fig. 1.10. The model for initial stages of recombination by Hotchkiss (1971). Interrupted lines denote newly-synthesized DNA. (A) A "poisoned arrow" – a 3'-ended single strand complexed by specialized recombinational enzymes is ready to invade the homologue. (B) The single strand formed a triplex with the homologous duplex, the 3'-end priming DNA synthesis. (C) The homologous strand of the duplex is broken as a result of this synthesis and goes to the other duplex, resulting in formation of a Holliday junction. (D) The invading strand is displaced and returns to the original duplex with the information copied from the homologous chromosome.

break, DNA synthesis is primed on the homologous chromosome. The elongated strand is then either withdrawn from the homologue, or induces displacement and breakage of the homologous strand in the homologue with subsequent formation of a Holliday junction. This model can be called "global break-join with local copy-choice". The current relevance of this model is in its predictions of the invasiveness of 3'-ends complexed with specialized enzymes and of the initial DNA triplex intermediate that precedes DNA synthesis.

The mid-seventies were marked by appearance of the first "true" models of recombinational repair that were focused on repair of DNA damage rather than on its genetic consequences. However, they went mostly unnoticed at the time of their appearance, like the first mammals among the dinosaurs — the reigning models for crossing-over and gene conversion. What really captured the minds at that time was a modification of the Holliday model, proposed by Meselson and Radding in 1975, which combined Holliday and Fox proposals together in an attractive scheme[50] (Fig. 1.11). Mechanistically, the Meselson-Radding model was not much different from the earlier scheme of Hotchkiss (1971), although it was nicely elaborated. Meselson-Radding's was the

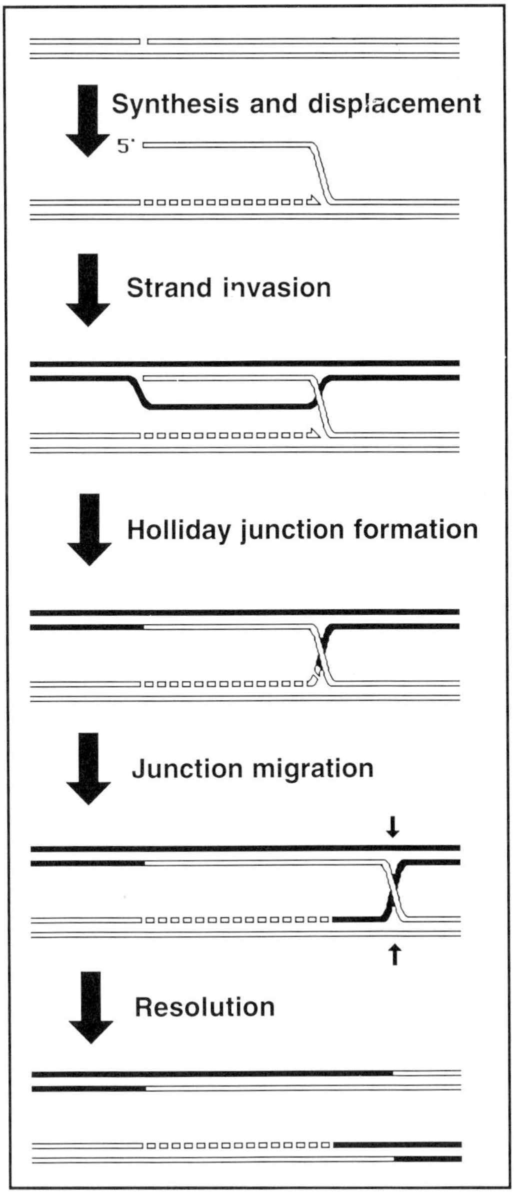

Fig. 1.11. The Meselson-Radding model of crossing-over (1975). The newly-synthesized DNA is shown as interrupted lines. The direction of Holliday junction resolution is indicated with small arrows. For clarity, only one resolution product (cross-over) is shown. Explanations in the figure and in the text.

last scheme designed solely to rationalize recombination as a phenomenon of its own. For example, it proposed that a damaged DNA molecule donates information to the intact homologue. In retrospect, it is seen as a strategic flaw which sealed its doom. After riding high for about ten years, this last recombinational dinosaur became extinct, and now its carcass has only a nostalgic value.

Perhaps the first true model of recombinational repair was proposed in 1974 by Skalka, who studied phage replication, and wished to account for the formation of DNA molecules replicating in sigma-mode.[51] First, she independently proposed, preceded by Hanawalt,[52] that a replication fork will disintegrate if it encounters a single-strand interruption in template DNA. This notion predicted another major class of DNA lesions — a solitary double-strand end. However, in contrast to Hanawalt, who thought that these lesions are lethal, Skalka proposed a mechanism for their repair using homologous recombination. Her model differs from the Fox-66 model and is similar to Hotchkiss' thinking in suggesting that the invasion of a 3' end primes DNA synthesis. However, Skalka went further and detailed how the invading double-stranded end could generate a new replication fork, though she had not dealt with the junction (Fig. 1.12). Since the model was designed to explain phage replication phenomenology, no familiar "recombination" predictions were made. This is probably the reason why this break-copy-type scheme is not usually mentioned among the models of homologous recombination, although it describes basically the same process, yet happening between two genetically identical DNA molecules.

In 1976, Resnick, foreshadowed by an earlier proposal,[46] offered a model for the repair of chromosomal double-strand breaks, based on the interactions of the broken chromosome with an intact homologous chromosome[53] (Fig. 1.13). His model is distinct and quite specific about the initial and the concluding events of recombinational repair. It is peculiar (and reminiscent of the earlier Boon and Zinder 1969 scheme (Fig. 1.9) in suggesting that one of the ends of a double-strand break invades and copies the intact homologue but then is withdrawn to anneal with the awaiting other end. Resnick made few specific recombination predictions, dooming his model to years of obscurity. Perhaps, further experiments will secure the place for the Resnik scheme in the pantheon of recombinational repair models and mechanisms.

The progress of genetic and biochemical research eventually led to the formulation of two modern models of recombinational repair. In 1981, West, Howard-Flanders and colleagues proposed a scheme for daughter-strand gap repair,[54,55] which now celebrates its 15th year without major changes (Fig. 1.14). The process described by this model will be considered in detail in chapter 4. In 1983, the current model of double-strand break repair was proposed by Szostak and colleagues[56] to explain peculiarities of meiotic crossing-over as well as repair of

DNA lesions (Fig. 1.15). Because of the tremendous influence of this model, "recombinational repair" has become almost synonymous with "double-strand break repair", which is somewhat misleading. Double-strand break repair is discussed in detail in chapter 5.

Looking at the contemporary models of recombinational repair and their immediate predecessors we see three common phases of such repair. The central phase during which a damaged DNA sequence is complexed with an intact homologous sequence to form a joint molecule is called "synapsis". Accordingly, the phase preparatory for the synapsis is called "presynaptic", while resolution of the joint molecules

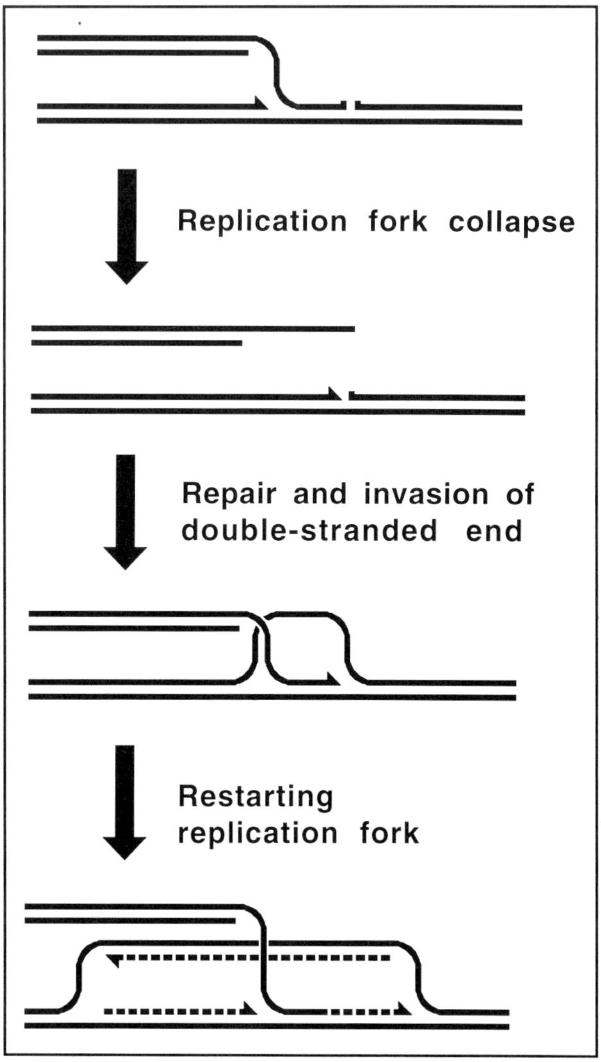

Fig. 1.12. Replication fork collapse and reinstallation by Skalka (1974). Newly-synthesized DNA is shown as interrupted lines. Explanations in the figure and in the text.

is referred to as the "postsynaptic" phase. This terminology was introduced by Clark some 25 years ago, at the very beginning of active exploration of recombinational repair.[57] We will use this convenient division while discussing details of recombinational repair pathways (chapters 4, 5, 6 and 7).

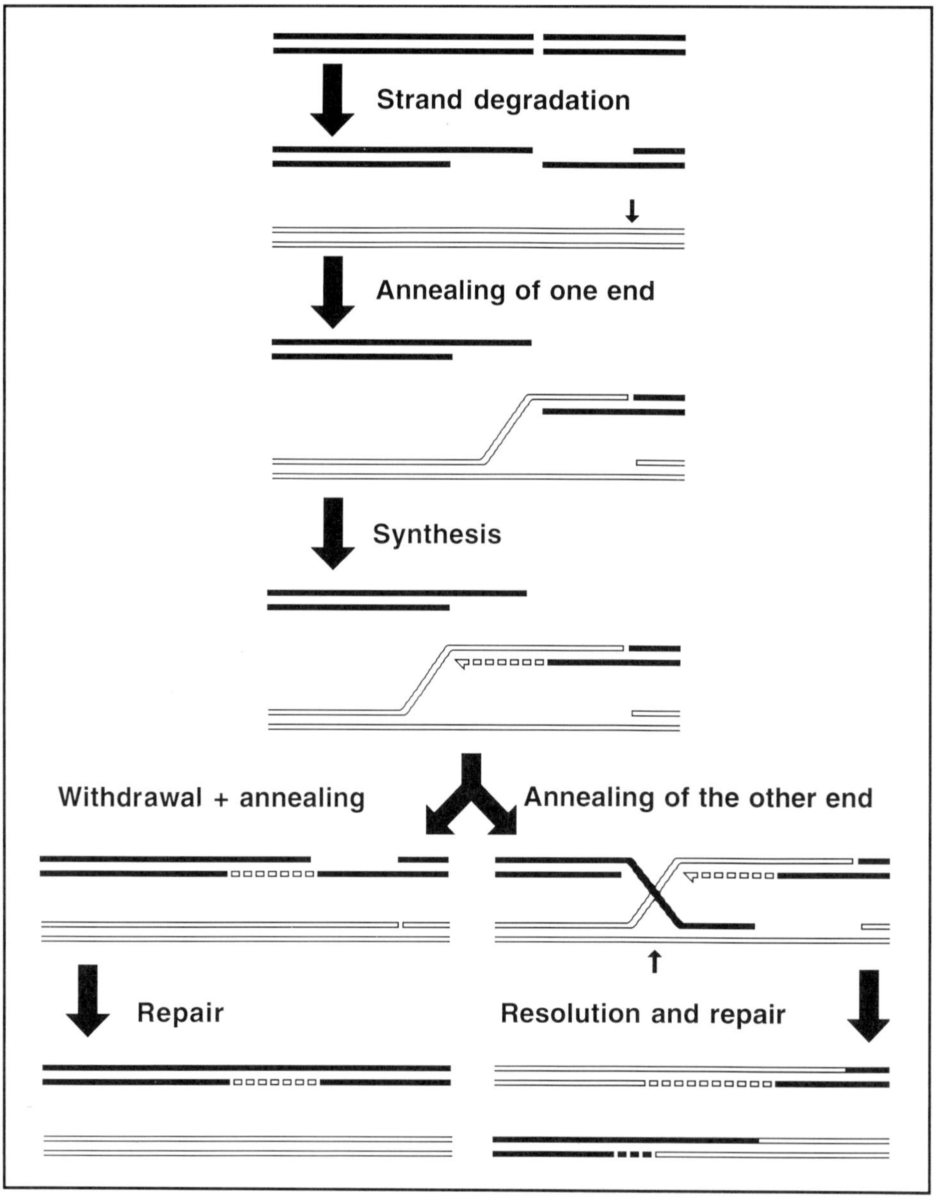

Fig. 1.13. Double-strand break repair according to Resnick (1976). Positions of single-strand scissions are indicated by small arrows, newly-synthesized DNA is shown as interrupted lines. Explanations in the figure and in the text.

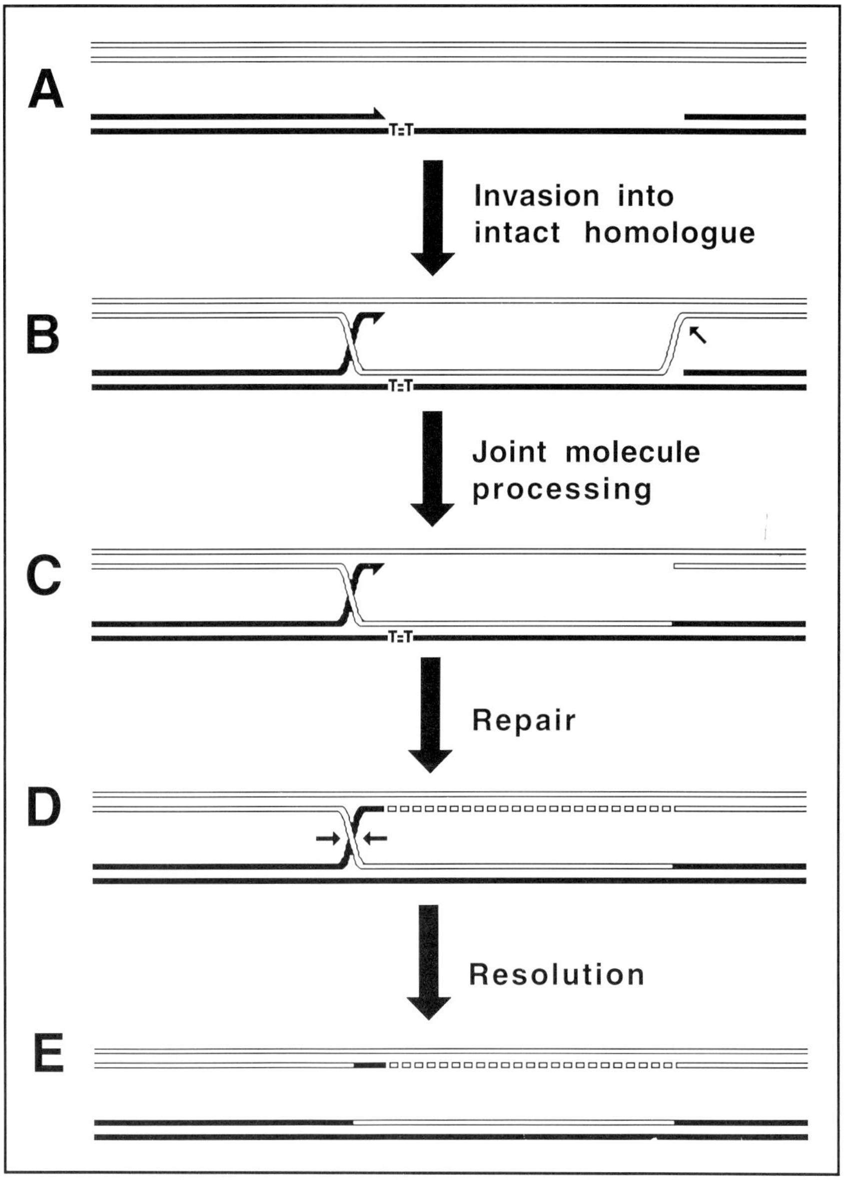

Fig. 1.14. A model for daughter-strand gap repair by West and others (1981). A thymine dimer is shown as T=T in the black duplex, positions of single-strand scissions are indicated by small arrows, newly-synthesized DNA is shown as interrupted lines. (A) DNA replication is blocked by a thymine dimer and reinitiates downstream, leaving a segment of template DNA adjacent to the dimer single-stranded. (B) The single-stranded gap invades a homologous duplex. (C) The three-strand junction at the right is removed by incision (the small arrow in "B"). (D) Thymine dimer in a duplex region is removed by the excision repair; the single-strand gap is filled by repair DNA synthesis. (E) The Holliday junction is resolved (the direction of the resolution is shown in "D"), freeing the participating duplexes from each other. For clarity, only one resolution product (noncross-over) is shown.

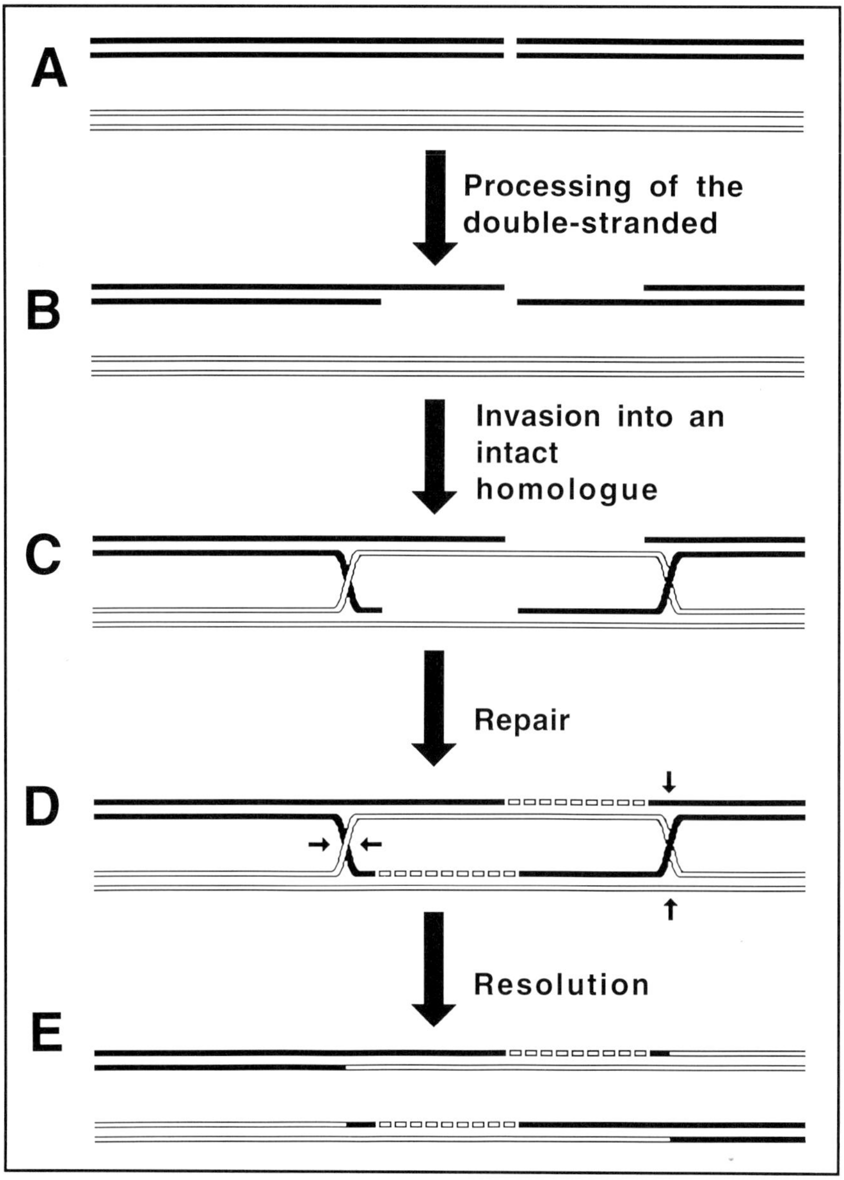

Fig. 1.15. The double-strand break repair model of Szostak and others (1983). (A) Two homologous DNA duplexes, the black one having a double-strand break, the white one intact. (B) The ends of the double-strand break are made single-stranded by exonucleases with the same polarity. (C) Resected ends invade an intact homologous chromosome. (D) The single-strand gaps are filled by repair synthesis. (E) The Holliday junctions are resolved (the direction of the resolution is shown in "D"), freeing the participating duplexes from each other. For clarity, only one (out of four) resolution product (cross-over) is shown.

1.2. WHY *E. COLI*?

Homologous recombination, a consequence of recombinational repair, is detected in almost any creature in which it is sought. Animals, insects, filamentous and unicellular fungi, — all enjoy genetic exchange.[58,59] Why, then, are the mechanisms of recombinational repair understood best for bacteria in general and for *Escherichia coli* in particular? A part of the answer is, of course, the ease of both biochemical and genetic manipulations in bacteria. The other part of the answer is that, while bacteria and lower eukaryotes use recombinational repair rather indiscriminately, higher eukaryotes continuously express genes of recombinational repair in a few tissues and restrict the scope of this repair in most somatic tissues.

1.2.1. The Nature of Two-Strand Lesions

Two-strand lesions appear in DNA in basically two different ways. Several types of lesions, such as double-strand breaks, cross-links, and broken replication forks are induced in DNA directly. Ionizing radiation (X- and gamma-rays), when passing through a solution, generates free radicals, which damage and break every molecule in their immediate vicinity. The energy deposition by gamma radiation allows formation of clusters of several radicals, so the damage to a big molecule near such a cluster can be significant.[60] Besides chemically modified bases and interruptions in one DNA strand, ionizing radiation also causes double-strand breaks. On the average, for each twenty single-strand breaks induced by X-rays in DNA, there is one double-strand break (reviewed in ref. 61). Another direct two-strand lesion, a cross-link, is observed in DNA treated with psoralen + UV-light or mitomycin C.[62,63] Finally, stalled replication forks are sometimes broken (see replication fork collapse below), releasing one of the replicating branches as a free double-strand end.[64]

An alternative source of two-strand lesions is replication of DNA with unrepaired one-strand lesions. In vitro, a noncoding lesion in template DNA (for example, an abasic site) blocks the progress of all three known *E. coli* DNA polymerases.[65] In vivo, replication is likely to reinitiate downstream, leaving an unfillable single-strand gap behind (Fig. 1.14). Such an unfillable gap is called a daughter-strand gap, since it appears in one of the two daughter branches after replication fork passage. Another one-strand lesion, a single-strand interruption in a template DNA, causes a disintegration (collapse) of a replication fork.[66] As a result, a double-strand end is detached from the full-length duplex molecule (Fig. 1.12).

1.2.2. Prokaryotes and Eukaryotes — Two Strategies of DNA Repair

The dissimilarity of organization and maintenance of the genome between higher eukaryotes and prokaryotes is likely to influence their needs for recombinational repair. First, the bulk of the cell mass in higher eukaryotes is represented by terminally-differentiated cells. Such cells require expression of only a tiny fraction of their genome for normal functioning, and so stability of DNA in general is not their first concern. Second, eukaryotic DNA is complexed with histones and is surrounded by a protein mass, which is likely to serve as a barrier between DNA and chemically-charged intracellular soup, reducing the amount of DNA damage. Third, double-strand breaks in DNA of higher eukaryotes can be repaired by simple ligation with a loss of several nucleotides at the break site without serious consequences for the cell. The coding sequences represent less than 30%, and sometimes as little as 1% of the genomes of higher eukaryotes,[67] making this cheap repair quite attractive. And fourth, one-strand lesions in DNA of higher eukaryotes are not only efficiently repaired, but also generate a signal for inhibition of DNA replication (7.0). This lowers the chances of formation of two-strand DNA lesions due to DNA replication on a damaged template. Consequently, most higher eukaryotic cells could do reasonably well without high capacity for recombinational repair. The only two cell types in higher eukaryotes which continuously express recombination repair genes are the cells of the immune system and the cells undergoing meiosis (7.2.1).

A very different situation is encountered in bacteria. First of all, every bacterial cell is an independent organism, struggling for survival in space and time amidst similar independent cells, which makes faithful propagation of the genomic DNA the primary concern of every cell. Second, naked bacterial DNA is in direct contact with the cytoplasm, unprotected from radicals and other reactive chemical species. Third, coding sequences in *E. coli* represent more than 95% of its genome which hardly makes a simple imprecise ligation an affordable option. And fourth, although prokaryotes have an efficient one-strand repair, they seem to lack mechanisms to block DNA replication when their chromosome is damaged (3.1.1). All this makes replication-induced two-strand DNA lesions a potential issue in bacteria, necessitating continuous expression of recombinational repair genes in every bacterial cell. This is the other part of the reason why recombinational repair has been studied mostly in bacteria in general and in *E. coli* in particular. Two of the above features, unicellular existence and the lack of dispensable sequences, are also applicable to lower eukaryotes like yeast making these organisms favorites for studies of recombinational repair mechanisms in eukaryotes.

1.2.3. The Recombinational Repair Capacity of Bacterial Cells

Wild type *E. coli* cells grown in a thin medium are able to survive some 60 cross-links.[68] It can be calculated on the basis of the data with excision-deficient strains[69] that *E. coli* survives about 100-200 daughter-strand gaps per chromosome. Bacterial cells should also be able to tolerate multiple collapses of replication forks, attaching the resulting double-stranded ends faithfully back to the circular domains of the chromosome. The only two-strand lesion proved to be deadly for *E. coli* is double-strand break. *E. coli* survives only 2-3 double-strand breaks in its chromosome,[70,71] suggesting that whenever a double-strand break occurs in unreplicated portion of the genome, it cannot be repaired. Whether *Escherichia coli* is an exception among bacteria in its inability to repair multiple double-strand breaks in the genome remains to be determined. There is a bacterium, *Deinococcus radiodurans*, which can repair >100 double-strand breaks per chromosome[72] — this unusual creature lives in conditions prohibitive for most other microbes; however, its extreme resistance to DNA damage is clearly an exception rather than a rule in bacterial world. Therefore, the poor capacity of *E. coli* and any other bacteria to repair double-strand breaks suggests that this type of DNA damage is infrequent in these organisms. Double-strand breaks excluded, two-strand lesions in *E. coli* either are the result of DNA replication on a damaged template (daughter-strand gaps, disintegrated replication forks) or require DNA replication for their repair (cross-links). In other words, recombinational repair acts to restart DNA replication inhibited at unrepaired one-strand lesions. This makes a short introduction into the mechanisms of DNA replication in *E. coli* a prerequisite for the future discussions of the mechanisms of recombinational repair in this organism.

1.3. DNA REPLICATION IN *E. COLI*

The entire 4.72 Mbp circular chromosome of *E. coli* is replicated by a single replication bubble emanating from the unique replication origin. Two replication forks meet each other in a chromosome region right across from the origin, which is called the terminus. The terminus is delineated by termination sites arranged so as to form a replication fork trap — replication forks can enter the terminus, but they cannot exit it.[73] Both replication forks of the replication bubble are active (replication is bidirectional). To replicate the whole chromosome within a short bacterial cell cycle (less than an hour), replication forks have to proceed with a neck-breaking speed of about 1 kb per second. However, even this speed is insufficient when the cell cycle of *E. coli* is squeezed to 24 minutes in a very rich medium. To prevent under-replication, *E. coli* starts a new round of DNA replication well before the completion of the ongoing round. Thus, in cells growing in a rich medium, there are from one to three replication bubbles (two to six replication forks).[74]

We can enjoy a detailed description of the mechanisms of DNA replication in *E. coli*, a product of more than 30 years of intensive investigation (for the in-depth reviews see refs. 75, 76). An entire process of replication of a chromosome can be subdivided into three major parts: initiation, elongation and termination. Termination is the least understood part and is not immediately relevant to the needs of recombinational repair, although inhibition of DNA replication associated with termination might cause disintegration of replication forks with their subsequent recombinational repair.[64,77] Elongation phase of DNA replication is interesting in conjunction with the mechanisms of two-strand lesion formation as a result of DNA synthesis across a damaged template. Initiation phase is most interesting to us because it may provide an idea of how inhibited DNA replication might reinitiate after recombinational repair is completed.

1.3.1. Initiation of Chromosomal DNA Replication in *E. coli*

In order to start a replication fork, one needs first to open the DNA duplex. At the origin of chromosomal DNA replication, this opening is affected by binding of the initiator protein, DnaA, to its recognition sites. DnaA recognizes and binds a degenerate nonanucleotide (T/C)-(T/C)-(A/T/C)-T-(A/C)-C-(A/G)-(A/C/T)-(A/C).[78] At the origin of chromosomal DNA replication, four DnaA recognition sites are found in a cluster. Binding of 10-20 DnaA monomers to this cluster of DnaA binding sites leads to an opening of the DNA duplex in the nearby region.

In vivo, single-strand DNA is immediately complexed with single-strand DNA-binding protein (SSB), which precludes many other proteins from binding to ssDNA. In order to load DNA replication proteins onto SSB-complexed single-strand DNA, help from other proteins bound to neighboring duplex regions is needed. In the case of *E. coli* chromosomal origin, DnaA itself, still sitting on the adjacent duplex region, assists in this loading.

In contrast to RNA polymerases, DNA polymerases cannot start DNA synthesis on a ssDNA without a primer. The 10 nucleotide-long primers for *E. coli* DNA polymerase are laid by a special RNA polymerase called primase (DnaG protein). DnaG primase works in a complex with a DNA helicase (encoded by *dnaB* gene) that drives DNA unwinding at the replication fork. The complex of DnaG and DnaB proteins is called a mobile primosome; it propagates along the ssDNA in the 5'-to-3' direction, laying primers every 1.5-2.0 kb. The mechanics of primosome assembly at the origin of chromosomal DNA replication is as follows. While still in solution, DnaB protein is recruited by DnaC protein, which also has an affinity to DNA-bound DnaA protein. Hence, DnaC delivers DnaB helicase to single-stranded DNA if DnaA protein is bound nearby. When DnaB helicase is loaded onto single-stranded DNA and is associated with DnaG primase, the replicative primosome is formed.

The final stage of replication fork formation is the association of a multisubunit DNA polymerase III with the nascent replication bubble.

First, a DnaN protein dimer is clamped around a primed segment of DNA to form a ring that can move freely along the double-stranded DNA. DnaN clamp is called the processivity subunit of DNA polymerase III because it ensures that DNA polymerase stays bound to DNA during polymerization. DNA synthesis begins when DNA polymerase holoenzyme is loaded onto the DnaN clamp at the primer. There are up to 300 DnaN monomers per cell,[79] — some 10-fold excess of DnaN dimers over DNA pol III holoenzyme, which is present in 10-20 copies per cell.[80]

1.3.2. Elongation Phase of DNA Replication in *E. coli*

In an established replication fork, DnaB helicase (maybe with the help of auxiliary helicases like Rep and UvrD) unwinds parental duplex DNA, while the associated DnaG primase lays primers for both the leading and the lagging strands. DnaN clamps are formed around any new primer, while the single-stranded regions between primers are complexed with SSB. When the stretch of DNA between the two adjacent primers is duplicated (SSB is apparently displaced), DNA polymerase III is transferred from the used DnaN ring to a new DnaN ring, awaiting on the next primer (this explains the requirement for the excess of DnaN subunit over the holoenzyme). The two adjacent newly-synthesized DNA stretches, called Okazaki fragments, are separated by a single-strand interruption between the 3'-side of one of the fragments and the RNA primer, attached to the 5'-side of the other fragment. A one-subunit repair DNA polymerase (DNA pol I) starts DNA synthesis from the 3'-side of the interruption, simultaneously degrading the downstream RNA primer with its unique 5'-to-3' exonuclease activity. After the complete removal of the RNA primer, the single-strand interruption is sealed by DNA ligase.

This description corresponds to the behavior of DNA polymerase III during the lagging strand DNA synthesis in vitro. The overwhelming evidence indicates that in vitro, the leading strand DNA synthesis is continuous, so DNA polymerase III is loaded only once and then is able to replicate megabases of DNA before dissociation.[76] It is said that in vitro, the processivity of the leading strand DNA synthesis is greater than that of the lagging strand DNA synthesis. If DNA synthesis on the leading and the lagging strands has different processivity in vivo as well, the distribution of lengths of daughter-strand gaps (1.2.1), produced during replication of templates with lesions, would be bimodal, since the gaps in the leading strand would be much longer than the gaps in the lagging strand. However, the distribution of length of the daughter-strand gaps is unimodal, suggesting that the processivity of DNA synthesis in vivo is comparable for both strands.[81] Indeed, direct measurements of the initial products of DNA synthesis in vivo, which are possible if DNA ligase or DNA pol I are inactivated, unequivocally show that replication in *E. coli* is discontinuous on both strands.[82-85]

1.3.3. Initiation of Plasmid DNA Replication

Small multicopy plasmids of *E. coli* initiate replication of their DNA in a way very different from that of the chromosome. DNA at their replication origins is first transcribed, and a portion of the resulting several hundred nucleotide-long transcript forms a stable RNA-DNA hybrid with its template, displacing the complementary DNA strand into a so-called R-loop. Then the RNA portion of the hybrid is cleaved at a specific site to provide a primer for a limited DNA synthesis carried out by a repair enzyme, DNA polymerase I.

The complementary DNA strand, displaced as a result of this DNA pol I-catalyzed synthesis, is complexed by SSB. For this reason it would be inert for any further transaction if not for a primosome assembly site (PAS) it encompasses. Generally, PAS sites have a short region of dyad symmetry, surrounded by a particular DNA sequence. The region of dyad symmetry folds back on itself, forming a short duplex stretch, which can not be disrupted by SSB. Depending on the sequence of the duplex region and the surrounding context, a specific protein binds to it. If the hairpin contains a DnaA box, then it will be bound by DnaA protein, which would then promote the primosome assembly in a way similar to that at the chromosomal origin.[86] However, in their replication, plasmids rely mostly on a different PAS which is bound by PriA helicase. The primosome assembled by PriA has a different composition. First, for DnaB-DnaC delivery to single-stranded DNA associated with PriA, DnaT protein is used. Second, DnaB loading is assisted by PriB and PriC proteins. Third, PriB protein seems to stay associated with the final PriA-DnaB complex.[87] Only then DnaG primase associates with DnaB to lay primers. The PriA-containing primosome can move along ssDNA in both directions.

We went to such a great length describing replication initiation in small *E. coli* plasmids only because their priming mechanism is uniquely suitable for the needs of recombinational repair. As will be discussed in detail in the next chapter, the central reaction of recombinational repair is an invasion of a single DNA strand into a homologous duplex DNA. As a consequence of this invasion, the complementary strand of the duplex is displaced forming a D-loop. In the D-loop, the otherwise silent PAS sites could be unraveled permitting primosome assembly. In this way, a collapsed replication fork is not only put together again, but can be fully restarted. The critical role for the PriA-dependent primosome in the restart of inhibited replication forks was proposed.[88]

SUMMARY

One-strand DNA lesions are repaired within the same DNA molecule mostly by excision repair. Right from the very beginning, one-strand DNA repair was amenable to studies with biochemical methods. Two-strand DNA lesions need an intact DNA sequence (usually a sister chromosome) for repair. The repair of two-strand DNA lesions often generates a new (recombinant) chromosome if the intact chromosome was genetically different from the damaged one. Because of this association with genetic recombination, two-strand repair is known as recombinational repair. For a long time, recombinational repair has been studied indirectly, by genetic methods, as homologous recombination.

The three basic mechanisms of recombinational repair are: (1) daughter-strand gap repair; (2) double-strand break repair and (3) single-strand annealing repair. They correspond to two structural types of DNA lesions: unfillable single-strand gaps (1) and double-strand breaks (2,3). According to its origin, two-strand DNA damage could be subdivided into direct lesions (double-strand breaks, cross-links) and postreplicative lesions (daughter-strand gaps, disintegrated replication forks). The bulk of two-strand lesions are of postreplicative nature, that is, they are generated as a result of DNA replication on a template DNA containing unrepaired one-strand lesions.

Recombinational repair is studied mostly in prokaryotes and lower eukaryotes, like yeast. The DNA packaging, cell cycle control and terminal differentiation of somatic cells in higher eukaryotes restrict the scope of recombinational repair in these cells rendering it difficult to study. In contrast, unicellular organisms use recombinational repair to mend a wider variety of DNA lesions and offer both genetic and biochemical possibilities to study this process.

All the recombinational repair reactions can be subdivided into three distinct phases, presynapsis, synapsis and postsynapsis. The central, synaptic phase is common to the major recombinational repair reactions and is catalyzed by enzymes called synaptases. The next chapter is devoted to the unusual properties of synaptases.

References

1. Gilbert W (1986) The RNA world. Nature 319:618.
2. Orgel LE and Crick FHC (1993) Anticipating an RNA world. Some past speculations on the origin of life: where are they today? FASEB J 7:238-239.
3. Sancar A (1994) Structure and function of DNA photolyase. Biochemistry 33:2-9.
4. Shevell DE, Friedman BM and Walker GC (1990) Resistance to alkylation damage in *Escherichia coli*: Role of the Ada protein in induction of the adaptive response. Mutat Res 233:53-72.
5. Dianov G, Price A and Lindahl T (1992) Generation of single-nucleotide repair patches following excision of uracil residues from DNA. Mol Cell Biol 12:1605-1612.
6. Sancar A (1994) Mechanisms of DNA excision repair. Science 266: 1954-1956.
7. Modrich P (1994) Mismatch repair, genetic stability, and cancer. Science 266:1959-1960.

8. Howard-Flanders P (1973) DNA repair and recombination. Brit Med Bull 29:226-235.
9. Krasin F and Hutchinson F (1977) Repair of DNA double-strand breaks in *Escherichia coli*, which requires *recA* function and the presence of duplicate genome. J Mol Biol 116:81-98.
10. Luchnik AN, Glaser VM and Shestakov SV (1977) Repair of DNA double-strand breaks requires two homologous DNA duplexes. Mol Biol Reports 3:437-442.
11. Luchnik AN, Glaser VM, Soldatov SP and Shestakov SV (1979) On the mechanism of double-strand break repair in DNA of yeast, *Saccharomyces cerevisiae*. Dokl Acad Nauk USSR 244:213-216.
12. Haynes RH and Kunz BA (1981) DNA repair and mutagenesis in yeast. In: Strathern JN, Jones EW and Broach JR, ed(s). The Molecular Biology of the Yeast *Saccharomyces*. Cold Spring Harbor, N.Y.: Cold Spring Harbor Laboratory, pp 371-414.
13. Potter H and Dressler D (1979) DNA recombination: in vivo and in vitro studies. Cold Spring Harbor Symp Quant Biol 43:969-985.
14. Rückert J (1892) Zur entwickelungsgeschichte des ovarialeies bei selachiern. Anatomischer Anzeiger 7:107-158.
15. Morgan TH and Cattell E (1912) Data for the study of sex-linked inheritance in Drosophila. J Exp Zool 13:79-101.
16. Sturtevant AH (1913) The linear arrangement of six sex-linked factors in *Drosophila*, as shown by their mode of association. J Exp Zool 14:43-59.
17. Creighton HB and McClintock B (1931) A correlation of cytological and genetical crossing-over in *Zea mays*. Proc Natl Acad Sci USA 17:492-497.
18. Goldway M, Sherman A, Zenvirth D, Arbel T and Simchen G (1993) A short chromosomal region with major roles in yeast chromosome *III* meiotic disjunction, recombination and double-strand breaks. Genetics 133:159-169.
19. Harm W (1964) On the control of UV-sensitivity of phage T4 by the gene *x*. Mutat Res 1:344-354.
20. Clark AJ and Margulies AD (1965) Isolation and characterization of recombination-deficient mutants of *Escherichia coli* K-12. Proc Natl Acad Sci USA 53:451-459.
21. Howard-Flanders P and Theriot L (1966) Mutants of *Escherichia coli* K-12 defective in DNA repair and in genetic recombination. Genetics 53:1137-1150.
22. Rupp WD, Wilde III CE, Reno DL and Howard-Flanders P (1971) Exchanges between DNA strands in ultraviolet-irradiated *Escherichia coli*. J Mol Biol 61:25-44.
23. Cox MM (1991) The RecA protein as a recombinational repair system. Mol Microbiol 5:1295-1299.
24. Crow JF (1988) The importance of recombination. In: Michod RE and Levin BR, eds. The Evolution of Sex. An Examination of Current Ideas. Sunderland, Massachusetts: Sinauer Associates Inc., pp 56-73.

25. Smith JM (1988) The evolution of recombination. In: Michod RE and Levin BR, eds. The Evolution of Sex. An Examination of Current Ideas. Sunderland, Massachusetts: Sinauer Associates Inc., pp 106-125.
26. Bernstein H, Hopf FA and Michod RE (1988) Is meiotic recombination an adaptation for repairing DNA, producing genetic variation, or both? In: Michod RE and Levin BR, eds. The Evolution of Sex. An Examination of Current Ideas. Sunderland, Massachusetts: Sinauer Associates Inc., pp 139-160.
27. Janssens FA (1909) La théorie de la chiasmatypie. Nouvelle interpretation des cinèses de maturation. Cellule 22:387-411.
28. Belling J (1931) Chiasmas in flowering plants. Univ Calif Publ Botany 16:311-338.
29. Belling J (1931) Chromomeres of liliaceous plants. Univ Calif Publ Botany 16:153-170.
30. Hershey AD and Rotman R (1949) Genetic recombination between host-range and plaque-type mutants of bacteriophage in single bacterial cells. Genetics 34:44-71.
31. Lederberg J (1955) Recombination mechanisms in bacteria. J Cell Comp Physiol 45, Suppl. 2:75-107.
32. Watson JD and Crick FHC (1953) Molecular structure of nucleic acids. A structure for deoxyribose nucleic acid. Nature 171:737-738.
33. Watson JD and Crick FHC (1953) Genetical implications of the structure of deoxyribonucleic acid. Nature 171:964-967.
34. Meselson M and Stahl FW (1958) The replication of DNA in *Escherichia coli*. Proc Natl Acad Sci USA 44:671-682.
35. Meselson M and Weigle JJ (1961) Chromosome breakage accompanying genetic recombination in bacteriophage. Proc Natl Acad Sci USA 47:857-868.
36. Stahl FW (1956) The effects of the decay of incorporated radioactive phosphorus on the genome of bacteriophage T4. Virology 2:206-234.
37. Barricelli NA (1956) A "chromosomic" recombination theory for multiplicity reactivation in phages. Acta Biotheor 11:107-120.
38. Whitehouse HLK (1963) A theory of crossing-over by means of hybrid deoxyribonucleic acid. Nature 199:1034-1040.
39. Holliday R (1964) A mechanism for gene conversion in fungi. Genet Res 5:282-304.
40. Sigal N and Alberts B (1972) Genetic recombination: the nature of a crossed strand-exchange between two homologous DNA molecules. J Mol Biol 71:789-793.
41. Fox MS (1966) On the mechanism of integration of transforming deoxyribonucleate. J Gen Physiol 49:183-196.
42. Stahl FW, Kobayashi I and Stahl MM (1985) In phage λ, *cos* is a recombinator in the Red pathway. J Mol Biol 181:199-209.
43. Thomas CA (1966) Recombination of DNA molecules. Progress Nucl Acid Res Mol Biol 5:315-337.

44. Thomas Jr. CA (1967) The recombination of DNA molecules. In: Quarton GC, Melnechuk T and Schmitt FO, eds. The Neurosciences. A study program. New York: The Rockefeller University Press, pp 162-182.
45. Cassuto E and Radding CM (1971) Mechanism for the action of λ exonuclease in genetic recombination. Nature New Biol 229:13-16.
46. Korol'ov VG and Grachova LM (1972) Induction of recombination by decay of P32 incorporated into one or both genomes of diploid *Saccharomyces cerevisiae.* Genetika 8, #8:111-120.
47. Boon T and Zinder ND (1969) A mechanism for genetic recombination generating one parent and one recombinant. Proc Natl Acad Sci USA 64:573-577.
48. Hotchkiss RD (1971) Toward a general theory of genetic recombination in DNA. Adv Genet 16:325-348.
49. Paszewski A (1970) Gene conversion: observations on the DNA hybrid models. Genet Res 15:55-64.
50. Meselson M and Radding CM (1975) A general model for genetic recombination. Proc Natl Acad Sci USA 72:358-361.
51. Skalka A (1974) A replicator's view of recombination (and repair). In: Grell RF, ed. Mechanisms in Recombination. New York: Plenum Press, pp 421-432.
52. Hanawalt PC (1966) The U.V. sensitivity of bacteria: its relation to the DNA replication cycle. Photochem Photobiol 5:1-12.
53. Resnick MA (1976) The repair of double-strand breaks in DNA: a model involving recombination. J Theor Biol 59:97-106.
54. West SC, Cassuto E and Howard-Flanders P (1981) Mechanism of *E. coli* RecA protein directed strand exchanges in post-replication repair of DNA. Nature 294:659-662.
55. Livneh Z and Lehman IR (1982) Recombinational bypass of pyrimidine dimers promoted by the RecA protein of *Escherichia coli.* Proc Natl Acad Sci USA 79:3171-3175.
56. Szostak JW, Orr-Weaver TL, Rothstein RJ and Stahl FW (1983) The double-strand-break repair model for recombination. Cell 33:25-35.
57. Clark AJ (1971) Toward a metabolic interpretation of genetic recombination of *E. coli* and its phages. Annu Rev Microbiol 25:437-464.
58. Whitehouse HLK (1982) Genetic Recombination. Understanding the Mechanisms. Chichester: John Wiley & Sons Ltd., 415 p.
59. Kucherlapati R and Smith GR, eds (1988) Genetic Recombination. Washington, D.C.: American Society for Microbiology, 731 p.
60. Ward JF (1988) DNA damage produced by ionizing radiation in mammalian cells: identities, mechanisms of formation, and reparability. Prog Nucl Acid Res Mol Biol 35:95-125.
61. Krisch RE, Flick MB and Trumbore CN (1991) Radiation chemical mechanisms of single- and double-strand break formation in irradiated SV40 DNA. Radiat Res 126:251-259.
62. Cimino GD, Gamper HB, Isaacs ST and Hearst JE (1985) Psoralens as photoactive probes of nucleic acid structure and function: organic chemistry, photochemistry, and biochemistry. Annu Rev Biochem 54:1151-1193.

63. Tomasz M, Lipman R, Chowdary D, Pawlak J, Verdine GL and Nakanishi K (1987) Isolation and structure of a covalent cross-link adduct between mitomycin C and DNA. Science 235:1204-1208.
64. Kuzminov A (1995) Instability of inhibited replication forks in *E. coli*. BioEssays 17:733-741.
65. Bonner CA, Stukenberg PT, Rajagopalan M, Eritja R, O'Donnel M, McEntee K, Echols H and Goodman MF (1992) Processive DNA synthesis by DNA polymerase II mediated by DNA polymerase III accessory proteins. J Biol Chem 267:11431-11438.
66. Kuzminov A (1995) Collapse and repair of replication forks in *Escherichia coli*. Mol Microbiol 16:373-384.
67. Cavalier-Smith T (1985) Eukaryote gene numbers, non-coding DNA and genome size. In: Cavalier-Smith T, ed. The Evolution of Genome Size. Chichester: John Wiley & Sons, pp 69-103.
68. Sinden RR and Cole RS (1978) Repair of cross-linked DNA and survival of *Escherichia coli* treated with psoralen and light: effects of mutations influencing genetic recombination and DNA metabolism. J Bacteriol 136:538-547.
69. Wang T-CV and Smith KC (1983) Mechanisms for *recF*-dependent and *recB*-dependent pathways of postreplication repair in UV-irradiated *Escherichia coli uvrB*. J Bacteriol 156:1093-1098.
70. Krisch RE, Krasin F and Sauri CJ (1976) DNA breakage, repair and lethality after ^{125}I decay in *rec*$^{+}$ and *recA* strains of *Escherichia coli*. Int J Radiat Biol 29:37-50.
71. Ulmer KM, Gomez RF and Sinskey AJ (1979) Ionizing radiation damage to the folded chromosome of *Escherichia coli* K-12: repair of double-strand breaks in deoxyribonucleic acid. J Bacteriol 138:486-491.
72. Minton KW (1994) DNA repair in the extremely radioresistant bacterium *Deinococcus radiodurans*. Mol Microbiol 13:9-15.
73. Hill TM (1992) Arrest of bacterial DNA replication. Annu Rev Microbiol 46:603-633.
74. Helmstetter CE (1987) Timing of synthetic activities in the cell cycle. In: Neidhardt FC, ed. *Escherichia coli* and *Salmonella tiphimurium*. Cellular and Molecular Biology. Washington, D.C.: American Society for Microbiology, pp 1594-1605.
75. Baker TA and Wickner SH (1992) Genetics and enzymology of DNA replication in *Escherichia coli*. Annu Rev Genet 26:447-477.
76. Marians KJ (1992) Prokaryotic DNA replication. Annu Rev Biochem 61:673-719.
77. Horiuchi T and Fujimura Y (1995) Recombinational rescue of the stalled DNA replication fork: a model based on analysis of an *Escherichia coli* strain with a chromosomal region difficult to replicate. J Bacteriol 177:783-791.
78. Schaefer C and Messer W (1991) DnaA protein/DNA interaction: Modulation of the recognition sequence. Mol Gen Genet 226:34-40.

79. Burgers PMJ, Kornberg A and Sakakibara Y (1981) The *dnaN* gene codes for the β subunit of DNA polymerase III holoenzyme of *Escherichia coli.* Proc Natl Acad Sci USA 78:5391-5395.
80. Wu YH, Franden MA, Hawker JR and McHenry CS (1984) Monoclonal antibodies specific for the α subunit of the *Escherichia coli* DNA polymerase III holoenzyme. J Biol Chem 259:12117-12122.
81. Wang T-CV and Chen S-H (1992) Similar-sized daughter-strand gaps are produced in the leading and the lagging strands of DNA in UV-irradiated *E. coli uvrA* cells. Biochem Biophys Res Com 184:1496-1503.
82. Okazaki R, Okazaki T, Sakabe K, Sugimoto K, Kainuma R, Sugino A and Iwatsuki N (1968) In vivo mechanism of DNA chain growth. Cold Spring Harbor Symp Quant Biol 33:129-143.
83. Pauling C and Hamm L (1969) Properties of a temperature-sensitive, radiation-sensitive mutant of *Escherichia coli.* II. DNA replication. Proc Natl Acad Sci USA 64:1195-1202.
84. Louarn J-M and Bird RE (1974) Size distribution and molecular polarity of newly replicated DNA in *Escherichia coli.* Proc Natl Acad Sci USA 71:329-333.
85. Wang T-CV and Smith KC (1989) Discontinuous DNA replication in a *lig-7* strain of *Escherichia coli* is not the result of mismatch repair, nucleotide-excision repair, or the base-excision repair of DNA uracil. Biochem Biophys Res Com 165:685-688.
86. Masai H, Nomura N and Arai K (1990) The ABC-primosome. A novel priming system employing DnaA, DnaB, DnaC, and primase on a hairpin containing a DnaA box sequence. J Biol Chem 265:15134-15144.
87. Allen GC Jr. and Kornberg A (1993) Assembly of the primosome of DNA replication in *Escherichia coli.* J Biol Chem 268:19204-19209.
88. Seufert W and Messer W (1986) Initiation of *Escherichia coli* minichromosome replication at *oriC* and at protein n' recognition sites. Two modes for initiating DNA synthesis in vitro. EMBO J 5:3401-3406.

CHAPTER 2

RecA Protein – Homologous Pairing Activity

In order for a damaged DNA to be repaired with the help of an intact homologous sequence these two DNAs need to (1) find each other among the millions of unrelated sequences and (2) pair in a way that would allow the compromised or the lacking information on the damaged sequence to be reestablished with a help of the intact sequence. In *E. coli*, these intricate reactions are catalyzed by a single relatively small enzyme called RecA protein. The size of RecA is only about 38 kDa, but the protein is able to find a perfect match to any DNA sequence of 10 nucleotides or longer among millions of unrelated sequences. Moreover, RecA protein synapses the two homologous sequences in such a way that the damaged sequence can now be mended by one-strand repair using its intact homologous sequence as a template. How can such a small protein accomplish such an extraordinary complex task? The power of RecA is in numbers — the protein works as a polymer, which can comprise hundreds of RecA monomers.

2.1. IN VIVO ROLES OF RecA PROTEIN IN *E. COLI* AND ITS DISTRIBUTION AMONG OTHER BACTERIA

recA mutations are unusually pleiotropic (for an early yet informative review see ref. 1). *recA* mutants have reduced viability, are very sensitive to DNA damage, nonmutable by DNA damage, are deficient in homologous recombination and cannot induce prophages in response to DNA damage. These various phenotypes stem from a single deficiency of *recA* cells, the inability to form an active RecA filament. No extragenic suppressors that would cancel all the phenotypes stemming from *recA* deficiency has been found yet, suggesting that *recA* is the only gene of its kind in the *E. coli* genome.

2.1.1. *RECA* GENE AND PROPERTIES OF *RECA* NULL MUTANTS

recA was the first *E. coli* gene affecting recombinational repair to be discovered.[2] *recA* is not a part of any operon, is surrounded by genes unrelated to DNA metabolism and has its own promoter and terminator.[3,4] Normally, *recA* expression maintains from 1,000-10,000 RecA monomers per cell.[5-8] RecA production is induced by damage to DNA, such as UV-irradiation or nalidixic acid treatment, to result in up to 100-fold increase in the amount of the protein.[5,6] *recA* cells are extremely sensitive to DNA damage; nevertheless, *recA* null mutants are viable, though they grow slower than WT cells. The slower growth rate of *recA* cultures is due to the continuous generation of dead cells, rather than because of growth defects. The fraction of dead cells in *recA* mutant cultures reaches 20-40%.[9] Many of the dead *recA* cells completely lack chromosomal DNA (about 10% of the total cell count); even bigger fraction of all the cells degrade one of their several chromosomes.[10,11] Up to 20% of the total DNA in *recA* mutant cultures is degraded at any given moment.[12] However, the decent viability of *recA* mutants indicates that the fatal DNA damage does not happen too often, at least in laboratory conditions.

2.1.2. CELLULAR PROCESSES DEPENDENT ON RECA

When bacterial DNA suffers massive damage, the cells react by increasing synthesis of some 20 proteins (RecA is among them) in the so-called "SOS response" (chapter 3). The expression of these genes is ordinarily repressed by LexA protein (3.1.2). The SOS response is activated when LexA repressor molecules are degraded. The induction of the SOS response is absolutely dependent on RecA protein. RecA protein is activated by damaged DNA, and the activated form of RecA catalyzes self-cleavage of LexA repressor (2.2.2).

The damage to bacterial DNA also causes prophage induction — a lytic development of latent bacteriophages. Normally, the lytic development is repressed by prophage-produced repressors. Similarly to LexA repressor, viral repressors cleave themselves in the presence of the activated RecA (2.2.2), allowing lytic development of latent bacteriophages. This mechanism is apparently meant to save bacteriophages out of a lethally-damaged cell.

RecA plays a pivotal role in recombinational repair of such two-strand DNA lesions as daughter-strand gaps, double-strand breaks and cross-links. For example, WT *E. coli* cells survive 53-71 cross-links per chromosome, while *recA* cells are killed by a single cross-link in their chromosome.[13] RecA-dependent mechanisms of recombinational repair are discussed in chapters 4 and 5.

Sometimes, a bacterial cell acquires an extra linear piece of chromosome of variable length from another cell (for the various mechanisms of such acquisition, see ref. 14). During conjugation, this piece is transferred from another live cell, which has a conjugative plasmid integrated into its chromosome. During transduction, this piece is delivered by a bacteriophage, whose capsid has mistakenly packaged a fragment of the host DNA instead of the phage chromosome. During transformation, a cell picks up a piece of DNA from the environment, from a dead decomposing cell. A WT cell treats this linear piece of homologous DNA as a chromosome that suffered double-strand breaks and repairs it off an intact homologue, i.e., its own chromosome. As a result, a recombinant chromosome is generated, in which parts of the exogenous DNA are combined with the endogenous sequences. It is not surprising that *recA* mutants are profoundly defective in all these types of chromosomal recombination (reviewed in ref. 15).

At last, two-strand DNA damage induces RecA-dependent mutagenesis. In general, DNA modification is mutagenic in that it causes the changes of single nucleotides in DNA sequences, which are called point mutations. Especially strong mutagens are those making DNA bases change their coding interactions. For example, guanine codes for cytosine in the opposite position. Oxidative damage of guanine makes it recognize adenine, which will be incorporated in a newly synthesized DNA across such a modified guanine, causing a point mutation.[16] However, RecA-dependent mutagenesis has a completely different nature. Some DNA modifications generate the so called noncoding lesions — bases that are completely missing or so distorted that they are not recognized by DNA polymerases. DNA replication can bypass such a noncoding lesion in a RecA-dependent reaction. During this by-pass, a DNA polymerase has to incorporate random nucleotides in the new DNA chain across from the damaged position, causing formation of mutations (4.6).

2.1.3. RecA Proteins in Other Bacteria

Many (but not all) bacterial RecA-like proteins can be cloned by complementation of recombinational repair defects of *E. coli recA* mutants, which facilitates their characterization. RecA proteins have been found in a variety of prokaryotes, including gram-negative as well as gram-positive bacteria, actinomycetes and cyanobacteria (reviewed in ref. 17). All these genes encode proteins with molecular weights 35-40 kDa (*E. coli* RecA is 38 kDa), which show high degree of similarity among each other and homology in regions believed to be critical for the synaptase function.[18] RecA proteins from *Bacillus subtilis* and *Deinococcus radiodurans* are poisonous for *E. coli*,[19,20] — they might be activated for recombination in the *E. coli* intracellular environment. The universal presence of RecA-like proteins among prokaryotes underscores the importance of their function for chromosomal DNA repair. *recA* gene is found even in the smallest known bacterial genome of *Mycoplasma* (its size is a mere 12% of the *E. coli's* genome), an intracellular parasite which is believed to have retained only the most necessary genes.[21]

2.2. IN VITRO ACTIVITIES OF RECA

Genetic studies demonstrated the central role of *recA* gene in recombinational repair of *E. coli*. However, the mechanisms of recombinational repair remained obscure until the purification of RecA protein and in vitro characterization of its unusual properties. RecA-promoted in vitro reactions can be summarized as invasion of single-stranded DNA into a homologous duplex DNA. The requirement for single-stranded DNA in RecA-promoted in vitro reactions suggests that in vivo two-strand DNA damage is always associated with stretches of single-stranded DNA.

2.2.1. RECA PROTEIN ASSEMBLES ON SSDNA INTO A HELICAL FILAMENT

When added to DNA in vitro, RecA protein assembles around it into a spiral filament (Fig. 2.1A). The parameters of the filament vary with the conditions used. In the absence of ATP, the so-called "collapsed" filament is formed,[22] which is considered to be an inactive form. In the presence of ATP (or nonhydrolysable ATP analog, ATPγS), the "extended" filament is formed, which is proficient in all RecA activities (Fig. 2.1A). Since ATP is always present in vivo in high concentrations (3.2.2), only the extended RecA filament will be considered here. Most of the measurements of extended RecA filaments were done in the presence of ATPγS, because of the greater stability of the filaments in the absence of hydrolysis of high energy cofactor; however, the parameters of ATP-containing filaments are very similar.[23] The width of ATP-containing RecA filament is about 10 nM (100 Å),[24,25] which is five times the width of duplex DNA. Atomic structure of RecA filament shows that there is enough space inside the filament to accommodate two interacting DNA molecules.[26] The ATPγS-containing filament has about 6 RecA monomers per 95 Å turn, each RecA monomer binding about 3 nucleotides of ssDNA.[27] At physiological pH, RecA filament does not readily assemble on dsDNA (it does so at lower pH), however, a filament assembled on ssDNA extends into a contiguous double-stranded region.[28,29] The stoichiometry of dsDNA binding is the same: one RecA monomer binds 3 bp, or a single 95 Å helical turn of RecA filament holds about 18 bp of dsDNA (axial spacing between adjacent base pairs 5.1 Å).[30,31] Since the axial spacing between adjacent base pairs in native DNA duplex is 3.4 Å, it is said that duplex DNA included into RecA filament is extended 1.5 times[24,32] (Fig. 2.1A). This extension is thought to facilitate homology recognition between two DNA molecules, captured by RecA filament.

The assembly of RecA filament on ssDNA has a fixed directionality: filaments grow from the 5'-end toward the 3'-end along the ssDNA.[33] The maximal rate of RecA filament assembly in vitro is 30-40 monomers per second.[33] Assuming DNA binding stoichiometry of one RecA monomer per 3 nucleotides, it means that a growing RecA filament

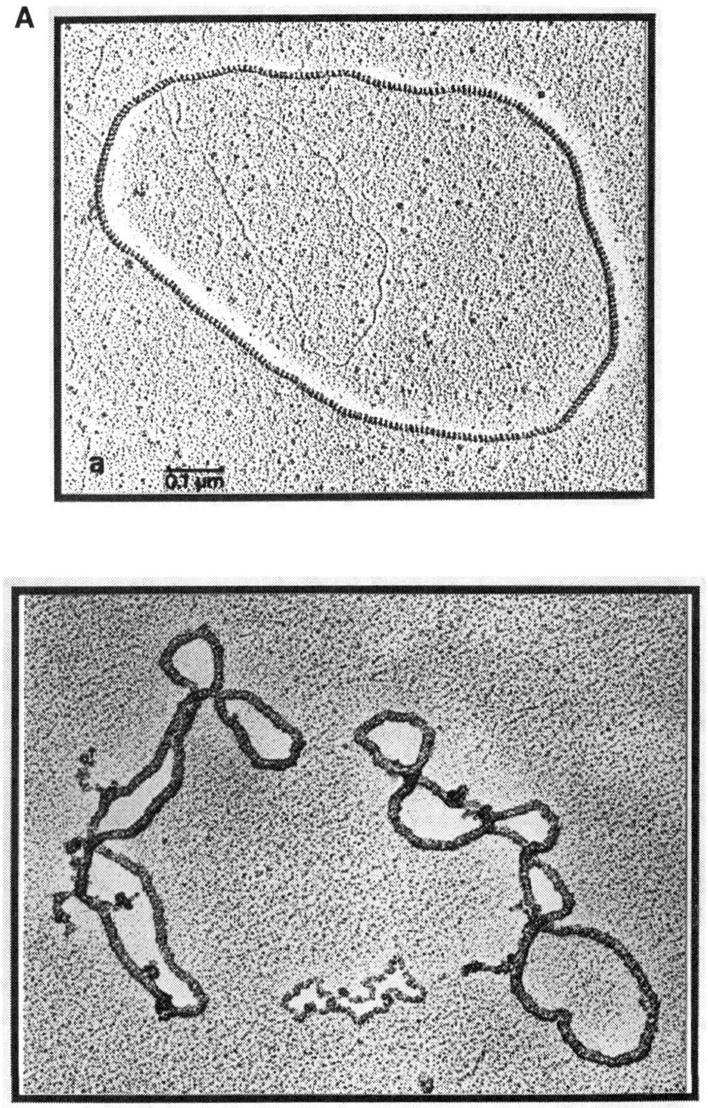

Fig. 2.1. RecA filaments on DNA compared with a naked DNA and ssDNA complexed by SSB. (A) Relaxed circular duplex DNA completely covered by RecA filament. Naked duplex DNA of the same length, lying mostly inside the RecA filament, illustrates that DNA within RecA filament is stretched 1.5 times. Reprinted with permission from Stasiak A et al, Nature 299:185-186. Copyright (1982) Macmillan Magazines Limited. (B) Two RecA-covered circular ssDNA molecules and one circular ssDNA of the same length complexed with SSB (in the middle). Note that while RecA filament stretches ssDNA, SSB compacts it. Reprinted with permission from Stasiak A et al, in Kucherlapati R et al, Genetic Recombination. Copyright (1988) American Society for Microbiology, pp 265-307.

engulfs about 100 nucleotides of ssDNA per second. In vitro, at the same time as the 3'-end of RecA filament keeps growing, the 5'-end begins disassembling, although at a slower rate, unveiling about 10 nucleotides per second.[29] In effect, the growing RecA filament treadmills along the DNA.

Every DNA-bound RecA monomer binds a single ATP molecule.[34] RecA filament assembled on ssDNA slowly hydrolyses ATP at a rate of about 30 ATP molecules per minute per monomer;[35] duplex DNA-bound filament has even lower ATPase activity.[29,36] However, the ATP hydrolysis is not needed for the assembly of filament on ssDNA, as the same extended RecA filament is formed in the presence of nonhydrolysable ATP analog, ATPγS.[25] Apparently, ATP serves as an allosteric cofactor for DNA binding in the extended conformation.[37] One role for ATP hydrolysis may be to promote filament disassembly since RecA filament formed in the presence of ATPγS does not disassemble on its own (see ref. 38 and references therein).

2.2.2. RecA Filament Cleaves LexA Repressor

Extended RecA filament promotes autocleavage of LexA repressor[39,40] as well as autocleavage of phage repressors[41-43] and of UmuD protein.[44] In these reactions, RecA plays a role of coprotease, since there are conditions under which LexA, phage lambda repressor and UmuD cleave themselves in the absence of RecA.[44,45] In vitro, efficient LexA cleavage is observed only when RecA filament is assembled on single-stranded DNA. Extended RecA filament assembled on duplex DNA promotes LexA cleavage at a 1/20th of the rate observed with a filament assembled on single-stranded DNA.[46] This feature of the SOS repressor cleavage makes biological sense — if all the single-stranded regions associated with DNA lesions are made double-stranded, there is no reason to further boost the cell's repair capacity.

The RecA-promoted LexA autocleavage is rapid and is independent of Mg^{2+} concentrations in the range of 1-10 mM.[47] In contrast, RecA-promoted autocleavage of UmuD and phage repressors is slow and observed only at Mg^{2+} concentrations of 10 mM or higher.[41-43] The possible biological significance of these differences will become clear later (3.2).

2.2.3. RecA Filament Finds a dsDNA Homologous to the "Inner" ssDNA

Having engulfed a ssDNA (which will be called "the inner ssDNA" henceforth) in the presence of ATP, RecA filament is able to search for a duplex DNA with homology to the inner ssDNA. The process should be very efficient, since the amount of nonhomologous DNA in vivo is overwhelming. In fact, an identical sequence, shifted even a single nucleotide forward or backward relative to the inner ssDNA, becomes perfectly nonhomologous to it. Another in vivo problem is a very high DNA concentration. If we assume the volume of a single *E. coli* cell to be 7.5 x 10^{-18} M^3 (cylinder 0.8 μM in diameter and 1.5 μM in length) and the cellular chromosomal content to be a single 4.7 x 10^6 bp genome (3.1 x 10^{-15} g), the DNA concentration in such a cell will be 410 μg/ml, which is, taking into consideration the molecular weight of such DNA, a solution of enormous viscosity. Moreover, since half of the cell volume is occupied by ribosomes, and there is usually more than one genome-equivalent of the chromosomal DNA per cell, the actual DNA concentration may be several times higher. In these conditions, RecA has to find a homology to the damaged DNA within minutes; otherwise, the damaged DNA will be degraded by nucleases or will bring about an even greater damage, as the upcoming replication fork will run into the damaged template.

The mechanism by which RecA filament searches for homology is still an enigma. An algorithm of homology search is likely to require repeated juxtaposition of a segment of the inner DNA with various segments of duplex DNA. If RecA filament is able to juxtapose two potentially nonhomologous sequences, how does it then let go a duplex which proved to be nonhomologous? One possible explanation is that a nonhomologous duplex is expelled out of the filament with the help of ATP hydrolysis. Indeed, as mentioned above (2.2.1), RecA filament continuously hydrolyzes ATP. However, at least in vitro, homology search goes unaltered in the presence of a nonhydrolysable ATP analog and, hence, does not require ATP hydrolysis.[48]

A clue to understanding the mechanism of RecA-catalyzed homology search is offered by a discrepancy in stoichiometries of RecA binding to single-stranded DNA, as measured by different methods. When a DNA molecule is asked how many RecA monomers have bound to it (by monitoring fluorescence of chemically-modified DNA or a fraction of DNA sedimenting with RecA filaments), the answer is one RecA monomer per six nucleotides.[49,50] In contrast, when RecA protein is asked the same question (by monitoring the rate of ssDNA-dependent ATP hydrolysis by RecA or by measuring RecA-DNA complex mass), the answer is one RecA monomer per three nucleotides (or base pairs of duplex DNA).[30,51] This disagreement of "the DNA story" and "the RecA story" indicates that after all the sites of RecA binding on the ssDNA are saturated, an additional equal amount of RecA can bind to such RecA-ssDNA filament and will be "unnoticed" by DNA.[52] The additional RecA monomers, although having apparently little or almost no contact with the inner DNA, are nevertheless fully capable of ATP hydrolysis and are able to bind another DNA molecule. Interestingly, it is this RecA-ssDNA complex which is proficient in subsequent reactions.

Lauder and Kowalczykowski[52] further investigated the possibility that RecA filament is a bipartite structure and found that there are indeed two distinct sub-populations of monomers in saturated RecA filaments. They confirmed and extended earlier observations showing that two subpopulations differ in respect to ATP and ssDNA binding: one population of RecA monomers binds both ATP and ssDNA 10 times as strong as the other one.[50,52,53] It was suggested that the presence of two subpopulations of monomers in RecA filaments could be important for homologous search.[52]

The idea of Lauder and Kowalczykowski about bipartite RecA filament could be developed into a model for RecA-catalyzed homologous search. We postulate that the subpopulation of "tight-binder" monomers forms a backbone of RecA filament and complexes the bulk of ssDNA, while the subpopulation of "loose-binder" monomers dissociates from the filament and reassociates back frequently. We further propose that only loose-binders can bind heterologous duplex DNA and try it against the inner ssDNA, held by the adjacent tight-binders. If the "inner" ssDNA does not recognize the juxtaposed segment of duplex DNA as homologous, the loose-binder holding this duplex dissociates from the filament, sets free the duplex and binds back to the filament. If, however, the inner ssDNA does recognize the duplex as homologous one, the two DNAs become paired within the stretch of the two tight-binder monomers adjacent to the successful loose-binder (Fig. 2.2).

The conclusion that a fraction of RecA monomers are cycled on and off the filament (which is the same as to say that they have 10 times lower affinity to DNA and ATP) solves the problem posed by the algorithm of homology search. This problem is best illustrated by an old observation that RecA filaments, assembled on a ssDNA, are "sticky": if mixed with duplex DNA, such filaments cause its transient precipitation.[54] The surprising part of this reaction is not the precipitation itself, which is expected since RecA filaments have to establish nonspecific contacts with other DNAs, but its transient nature. The model postulates that: (1) there are specialized monomers in RecA polymer (loose-binders) to bind "other" DNAs and (2) if "another" DNA proves to be nonhomologous to the inner DNA, the monomer that holds it has to dissociate from the filament. Thus, the fact that precipitates eventually dissolve would be attributed by the model to mass dissociation of loose-binder monomers from RecA filaments.

The idea that a single loose-binder monomer captures a segment of heterologous duplex DNA and tries it against the inner ssDNA rationalizes the fact that RecA can pair two sequences which share as few as 8 nucleotides of homology.[55] The functional division of labor between two populations of RecA monomers in filaments makes one wonder about the existence of creatures that might have two related but distinct RecA proteins — one a tight-binder, another a loose-binder. Bacteria, in general, seem to satisfy their recombinational repair needs with one species of RecA protein,[17] the examples of two distinct *recA* genes being extremely rare.[56] In contrast, yeast express at least three

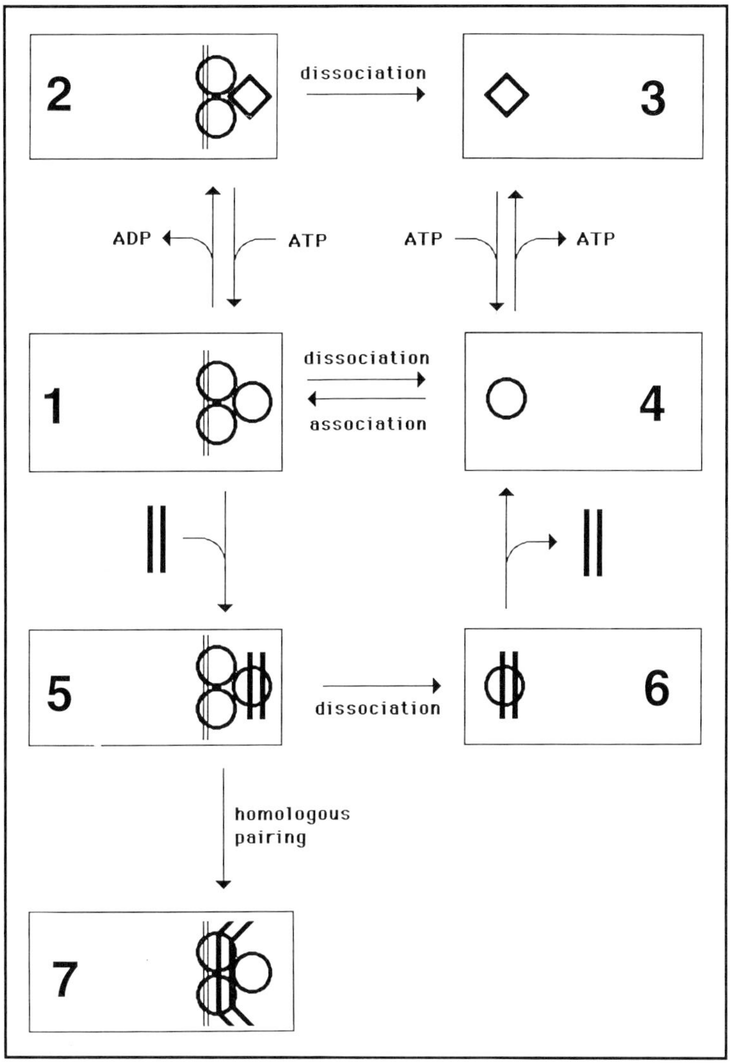

Fig. 2.2. The model of RecA filament functioning in ATP hydrolysis and during homology search. Circles, RecA monomers with ATP; diamonds, RecA monomers without cofactor, two thin lines – inner ssDNA, two thick lines – duplex DNA. 1. A segment of the filament with two tight-binder monomers holding to the ssDNA and one loose-binder monomer on the side. 2. The loose-binder hydrolyses ATP and dissociates the resulting ADP and phosphate (not shown). The loose-binder without cofactor either binds ATP (1) or leaves the filament (3). 4. RecA monomer without cofactor can bind ATP reversibly and after ATP binding can reassociate with the filament (1), again reversibly. 5. A loose-binder associated with filament binds duplex DNA. 6. If the duplex DNA is not recognized as homologous by the inner ssDNA, the loose-binder has to leave the filament and then it can dissociate from the duplex DNA (4). After that, the loose-binder can return back to the filament (1) or loose ATP (3). 7. If the duplex DNA is recognized as homologous by the inner ssDNA, the two DNAs are paired within the two tight-binders adjacent to the loose-binder.

RecA-like proteins during mitotic cell cycle, and induce an additional RecA-homolog during meiosis (7.1.1). It is tempting to speculate that in eukaryotes the division of labor is not only functional, but structural, with some proteins playing the role of tight-binders, while others act exclusively as loose-binders.

2.2.4. RecA Filament Catalyzes Exchange of Strands Between the "Inner" DNA and the Intact Homologue

When homology is found, that is, a homologous duplex DNA is aligned with the "inner" ssDNA, RecA filament catalyzes the exchange of strands between the two DNA molecules. In this process, the inner single-stranded molecule is paired with the complementary DNA strand from the intact duplex molecule, while the homologous strand of the duplex is displaced[57] (Fig. 2.3). The strand exchange is propagated along the RecA filament from the point of initial homologous contact in the 5'-to-3' direction,[58-60] i.e., in the direction of RecA filament assembly. The length of the DNA segments that trade strands in this process can be up to several kilobases.

If the inner DNA is a completely single-stranded molecule, the RecA-mediated strand exchange with a duplex DNA is called a "three-strand reaction". However, since RecA filament readily extends into the adjacent dsDNA regions, the initially three-strand exchange becomes a four-strand exchange in the region where the inner DNA becomes duplex. In this four-stranded reaction, there are two displaced strands, which are complementary to each other, so they anneal to form a second duplex, connected with the first one by crossed strands (Fig. 2.3).

The branch- or junction-migration, promoted by RecA is insensitive to mismatches and can overcome, although with decreased efficiency, numerous modified DNA bases and even deletions or duplications up to several hundred bases long.[61-63] ATP hydrolysis by RecA is not needed for the limited strand exchange between completely homologous sequences in a three-stranded reaction.[48] However, ATP hydrolysis increases the extent of strand exchange above 2 kb[64] and is needed to drive strand exchange in the four-stranded reaction.[65] ATP hydrolysis by RecA filament is absolutely required for the strand exchange to pass even a short heterology between two participating DNAs.[66,67]

2.2.5. The Role of SSB in RecA-Promoted Reactions

In physiological conditions, RecA binds only to single-stranded DNA. However, in vivo all the single-stranded DNA is likely to be complexed with single-stranded DNA-binding protein (SSB). Consequently, RecA and SSB might either compete for ssDNA binding (Fig. 2.1B) or somehow coexist on the same ssDNA. Peculiar patterns of SSB and RecA binding to ssDNA in different in vitro conditions will be discussed in

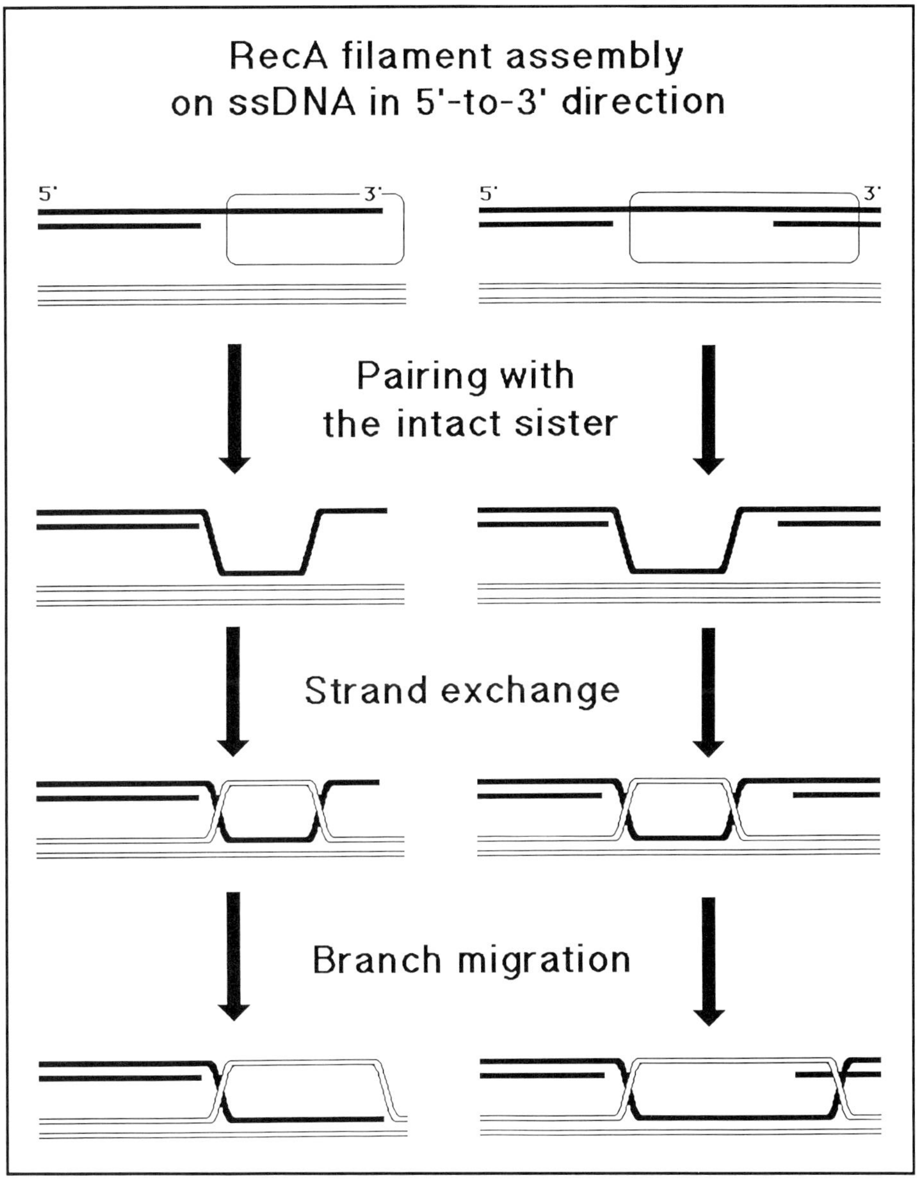

Fig. 2.3. The distinguishable phases of RecA-promoted reactions, as they are thought to happen in vivo. Black lines – damaged duplex, white lines – intact duplex, open rectangle with rounded corners – RecA filament. The left side – double-strand end repair; the right side – daughter-strand gap repair. Explanations in the figure.

detail in the next chapter (3.2). For now, it will suffice to say that in certain conditions SSB does not allow RecA onto ssDNA, whereas in other conditions, it is displaced by RecA or might coexist with RecA on the same ssDNA.

As was discussed above, RecA-promoted joint molecule formation can be divided into three distinct phases: (1) RecA filament polymerization on a single-stranded DNA; (2) homologous search and pairing of the inner ssDNA with homologous duplex DNA and (3) strand exchange, in which the inner ssDNA forms a new duplex with the complementary strand, while the homologous strand of the old duplex is displaced. SSB enhancement of in vitro RecA-mediated reactions is documented for the first and for the third phases.

In vitro, RecA is able to form long, strand exchange-proficient filaments on naked ssDNA only in low-salt low Mg^{2+} conditions which are far from being physiological. In fact, these conditions do not allow RecA filament to catalyze strand exchange itself. To stimulate strand exchange by preformed filaments, Mg^{2+} concentration needs to be raised. On the other hand, if RecA filaments are preformed at the elevated Mg^{2+} concentrations, the subsequent strand exchange is less productive. The explanation for this paradox is that in conditions which are closer to physiological ones, ssDNA readily forms secondary structures, which interfere with formation of long contiguous RecA filaments (Fig. 2.4). One way SSB enhances the performance of RecA is through elimination of these secondary structures, allowing formation of long contiguous RecA filaments at high Mg^{2+} concentrations.[68]

It is not clear to what extent SSB is displaced upon RecA binding to ssDNA. There is a report that SSB is not displaced from ssDNA by RecA filament, but seems to be included into ssDNA-RecA complex.[69] Such complexes contain one RecA monomer per four nucleotides, and one SSB monomer per 20-22 nucleotides, the values being 75% of saturation for both proteins. Nevertheless, these complexes are fully active in joint molecule formation.[69] Consistent with SSB retention within RecA filament is the way SSB enhances RecA performance by binding to the displaced strand during RecA-promoted strand-exchange. Potentially, the displaced strand can be utilized in a new round of strand-exchange reaction, serving as an "inner" DNA for a new RecA filament. The SSB binding sequesters the displaced strand from further action, making strand-exchange unidirectional.[70]

2.3. SUPERVISION OF RecA ACTIVITY

The efficiency of in vitro RecA-promoted pairing is alarmingly high. As mentioned above, RecA can pair two sequences that share a mere 8 nucleotides of homology (2.2.3); RecA can also promote extensive strand exchange between divergent DNA sequences (also called homeologous) (2.2.4). This quite indiscriminate nature of the RecA-promoted pairing poses a potential problem even in generally nonrepetitive genomes

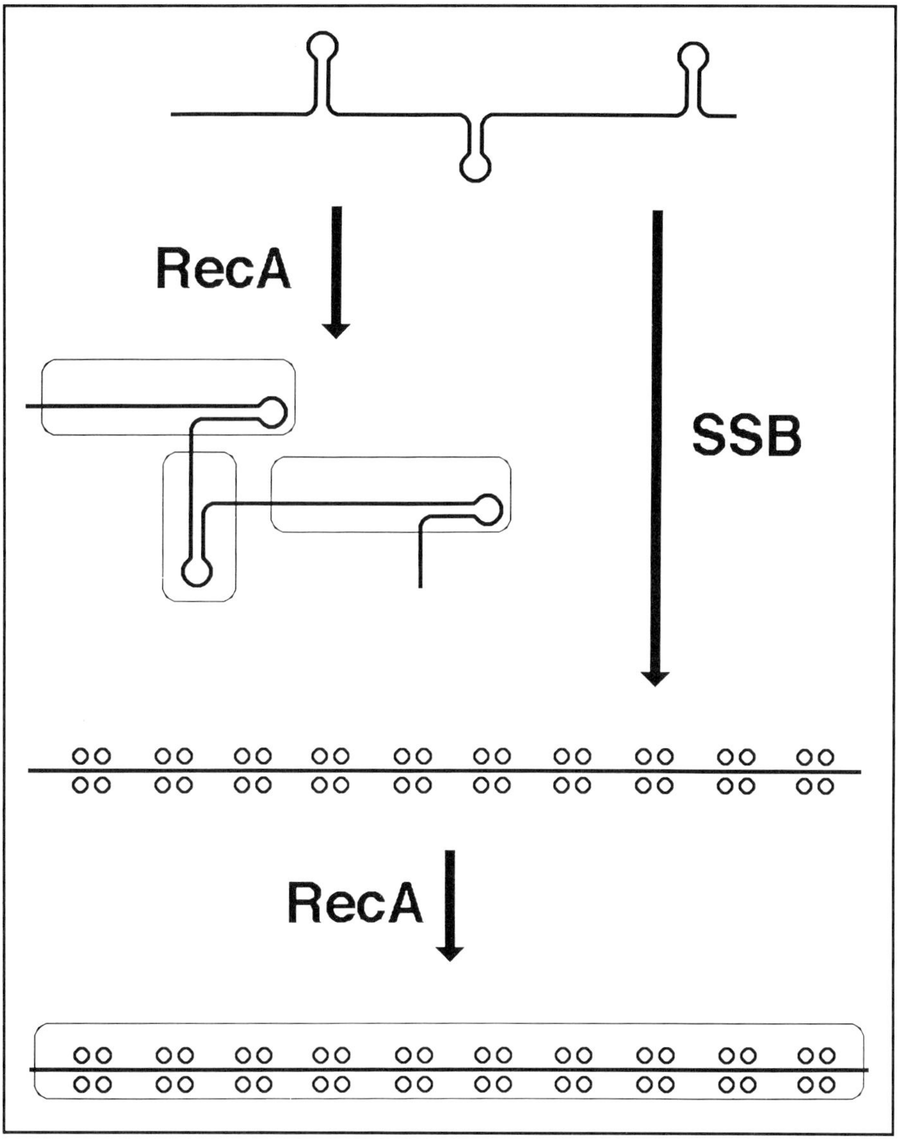

Fig. 2.4. At physiological Mg^{2+} concentrations, SSB is needed for assembly of functional RecA filament. Thick filled line – single-stranded DNA; stem-loop structures – secondary (duplex) structures in single-stranded DNA at physiological Mg^{2+} concentrations; open rectangles with rounded corners – RecA filaments; quartets of small circles – SSB tetramers. Without SSB (left side), formation of long contiguous RecA filament is compromised because the growth of nascent filaments is impossible past secondary structures. SSB helps RecA to polymerize into a long contiguous filament (right side) by ironing out secondary structures in the single-stranded DNA.

of bacteria. For example, in the *E. coli* genome, there are several ribosomal RNA operons which are mostly homologous to each other,[71,72] as well as many short (20-30 nucleotides) perfect repeats.[73] If not properly supervised, RecA could repair damage in one such repeat off another one. Such improper pairing, if accompanied by crossing-over, would lead to gross chromosomal rearrangements. The evidence is emerging that in *E. coli*, the components of the major mismatch-repair system make sure that RecA pairs a repeated sequence only with the identical sister sequence.

2.3.1. MutS and MutL Prevent Pairing Between Homologous Sequences

The supervision of the legitimacy of RecA-promoted pairing between long sequences is likely to be a secondary function of MutS and MutL proteins. These two proteins are central components of the major mismatch-repair pathway in *E. coli*. MutS binds to mismatches, while MutL is believed to transmit the signal of the MutS-mismatch interaction to the other parts of the correction system.[74] When presented with several copies of a repeated sequence, *mutS* and *mutL* mutant cells repair DNA damage in one repeat off a "wrong" repeat more frequently than WT cells. This is reflected in higher rates of recombination between homologous sequences in these two mutants.[75-78] Mechanistically, this phenomenon can be accounted for by the in vitro ability of MutS to inhibit RecA-promoted strand exchange between homologous sequences.[79] MutL enhances the efficiency of this inhibition. It is suggested that MutS binds to a newly-formed mismatch still within RecA filament, and that this binding inhibits further RecA-promoted strand exchange.[79] Thus, in the event of RecA catalyzing formation of a lengthy strand-exchange product containing several mismatches, completion of such a product in vivo is likely to be aborted by MutS and MutL.

2.3.2. Helicase II May Disrupt RecA-Mediated Pairing of Insufficient Length

The in vitro ability of RecA to pair ssDNA and a duplex DNA which have in common fewer than 10 contiguous nucleotides[55] spells another potential in vivo problem for recombinational repair. The term "homology" does not put the lower limit to what is still a homology, and what should be treated as a "coincidence". How does RecA discriminate against a 10 nucleotide homology in favor of a long, genuine homologous sequence?

The answer may be that it does not. UvsX protein is a RecA-like protein of bacteriophage T4. When catalyzing a homologous pairing reaction in vitro, UvsX protein tangles substrate DNAs into an awful knot and seems to be unable to complete strand exchange.[80] To assist UvsX protein in catalyzing complete strand exchange, T4 has a specific protein, Dda helicase. Dda helicase does two seemingly opposing things: it promotes the formation of the products of the UvsX-catalyzed strand exchange reaction,[81] and at the same time dissociates a lot of reaction intermediates back to initial substrates.[82] Its activity can be rationalized by assuming that Dda drives the correct interactions between substrates to completion while aborting incorrect or multiple interactions.

A similar situation could exist in *E. coli.* RecA filament can find a homology, but may have a difficult time assessing how relevant this sometimes limited homology is to the repair needs of the cell. *E. coli* could supervise this aspect of RecA activity with its own Dda-like protein — helicase II, coded by *uvrD* gene. Helicase II has a high copy number, 5,000-8,000 per cell,[83] which is further elevated during SOS induction (3.1.2). Helicase II unwinds DNA from nicks or double-strand ends.[84] Its role in the excision repair and the methyl-directed mismatch repair (two major types of one-strand repair (1.1.1)) is to remove the single-stranded DNA stretches which are to be resynthesized.[74] Like the discussed above *mutS* and *mutL* mutants, *uvrD* mutants exhibit a "hyper-recombinational" phenotype, although in a different setup. *uvrD* mutants are not particularly hyper-rec in systems based on a lengthy strand exchange between homeologous chromosomes.[76,78] However, they are hyper-rec when the strand exchange is expected to be limited in length.[75,85,86] In vitro, when added to a RecA-mediated strand-exchange reaction, UvrD helicase promotes both the completion of RecA-mediated strand-exchange and disassembly of strand exchange intermediates back to the initial substrates.[87] This is reminiscent of the mentioned above in vitro behavior of phage T4 Dda helicase and is consistent with the proposed role for UvrD as a protein that increases the fidelity of RecA-promoted pairing by disrupting homologous contacts of insufficient length.

Appendix: Importance of phage T4 for studies of recombinational repair in *E. coli*

Recombinational repair plays a pivotal role in the life cycle of T4, a big virulent bacteriophage of *E. coli,* by providing primers for its late DNA replication (reviewed in ref. 88). T4 has its own RecA-like protein, UvsX;[89-92] in fact, a *uvsX* mutant (simply *x* mutant at that time) was the first synaptase mutant to be discovered[93] — one year ahead of *E. coli recA* mutants. From the very beginning, the studies of T4 and *E. coli* machineries of recombinational repair developed in parallel and were mutually beneficial. Several important findings were made first in T4 and only then in *E. coli.* The idea of recombinational repair was first spelled out based on the finding that UV-resistance in T4 is partly dependent on the gene *x,* also responsible for homologous recombination in this phage.[93] In fact, the very notion of recombinational repair logically follows from a phenomenon called "multiplicity reactivation", first described for the same T4 and its relatives.[94] Multiplicity reactivation is the ability of several phage particles with lethally-damaged genomes to produce a progeny upon infecting a single cell. Several other processes, like involvement of a helicase in the supervision of RecA-catalyzed homologous pairing, mentioned above, are more pronounced in T4, and so have been also initially described in the phage system. One more example is the sequestration of the displaced strand by SSB during RecA-promoted strand exchange, which also was first suggested for T4 gp32 (SSB analog of T4).[95] There are other instances in this book when we will be referring to a particular T4 recombinational repair reaction because of its immediate relevance to analogous processes in *E. coli*.

SUMMARY

Synapsis of damaged DNA sequences with intact homologous sequences is catalyzed by enzymes called synaptases. RecA protein of *E. coli* is the most extensively studied synaptase. *recA* mutants are deficient in a variety of repair or repair-related processes and have a decreased viability. In vitro, RecA protein polymerizes into a helical filament around DNA starting from single-stranded regions. RecA filament assembled on single-stranded DNA promotes autocleavage of LexA (the repressor of the SOS response), UmuD (the regulator of translesion DNA synthesis) and phage repressors. RecA filament finds a duplex DNA sequence homologous to the "inner" DNA strand and catalyzes strand exchange — formation of a new duplex comprising the "inner" strand and its complement of the old duplex, the other strand of the old duplex being displaced. In order for RecA-promoted in vitro reactions to be efficient, "inner" DNA must be at least partially single-stranded, suggesting that in vivo, two-strand DNA damage is always associated with stretches of single-stranded DNA.

Since RecA-catalyzed homologous pairing tends to be indiscriminate, it is supervised in vivo at two levels. RecA-mediated strand exchange between homologous DNAs is aborted by MutS and MutL proteins, while synapsis between microhomologies is likely to be recalled by UvrD helicase.

Single-strand DNA binding protein (SSB) helps in RecA-promoted reactions in vitro by assisting with RecA polymerization and by making strand-exchange unidirectional. However, the in vivo interaction of SSB and RecA is likely to be more complex, since in certain in vitro conditions SSB precludes RecA binding to ssDNA. The next chapter will consider how *E. coli* could regulate its response to DNA damage using the balance of the SSB versus RecA competition for ssDNA. The bacterial strategy for dealing with replication of a damaged chromosome, called the SOS response, will be discussed in detail.

References

1. Clark AJ (1973) Recombination deficient mutants of *E. coli* and other bacteria. Annu Rev Genet 7:67-86.
2. Clark AJ and Margulies AD (1965) Isolation and characterization of recombination-deficient mutants of *E. coli* K-12. Proc Natl Acad Sci USA 53:451-459.
3. Horii T, Ogawa T and Ogawa H (1980) Organization of the *recA* gene of *Escherichia coli*. Proc Natl Acad Sci USA 77:313-317.
4. Sancar A, Stachelek C, Konigsberg W and Rupp WD (1980) Sequence of the *recA* gene and protein. Proc Natl Acad Sci USA 77:2611-2615.
5. Karu AE and Belk ED (1982) Induction of *E. coli recA* protein via *recBC* and alternative pathways: quantitation by enzyme-linked immunosorbent assay (ELISA). Mol Gen Genet 185:275-282.
6. Salles B and Paoletti C (1983) Control of UV induction of RecA protein. Proc Natl Acad Sci USA 80:65-69.
7. Moreau PL (1987) Effects of overproduction of single-stranded DNA-binding protein on RecA protein-dependent processes in *Escherichia coli*. J Mol Biol 194:621-634.
8. Sassanfar M and Roberts JW (1990) Nature of the SOS-inducing signal in *Escherichia coli*. The involvement of DNA replication. J Mol Biol 212:79-96.
9. Capaldo FN, Ramsey G and Barbour SD (1974) Analysis of the growth of recombination-deficient strains of *Escherichia coli* K-12. J Bacteriol 118:242-249.
10. Capaldo FN and Barbour SD (1975) DNA content, synthesis and integrity in dividing and non-dividing cells of Rec^- strains of *Escherichia coli* K12. J Mol Biol 91:53-66.
11. Skarstad K and Boye E (1993) Degradation of individual chromosomes in *recA* mutants of *Escherichia coli*. J Bacteriol 175:5505-5509.
12. Clark AJ, Chamberlin M, Boyce R and Howard-Flanders P (1966) Abnormal metabolic response to ultraviolet light of a recombination deficient mutant of *Escherichia coli* K12. J Mol Biol 19:442-454.
13. Sinden RR and Cole RS (1978) Repair of cross-linked DNA and survival of *Escherichia coli* treated with psoralen and light: effects of mutations influencing genetic recombination and DNA metabolism. J Bacteriol 136:538-547.
14. Amabile-Cuevas CF (1993) Origin, evolution and spread of antibiotic resistance genes. Austin: R.G. Landes Company, 102 p.
15. Mahajan SK (1988) Pathways of homologous recombination in *Escherichia coli*. In: Kucherlapati R and Smith GR, eds. Genetic Recombination. Washington, D.C.: American Society for Microbiology, pp 87-140.
16. Michaels ML and Miller JH (1992) The GO system protects organisms from the mutagenic effect of the spontaneous lesion 8-hydroxoguanine (7,8-dihydro-8-oxoguanine). J Bacteriol 174:6321-6325.
17. Miller RV and Kokjohn TA (1990) General microbilogy of *recA*: environmental and evolutionary significance. Annu Rev Microbiol 44:365-394.

18. De Mot R, Laeremans T, Schoofs G and Vanderleyden J (1993) Characterization of the *recA* gene from *Pseudomonas fluorescens* OE 28.3 and construction of a *recA* mutant. J Gen Microbiol 139:49-57.
19. Cheo DL, Bayles KW and Yasbin RE (1992) Molecular characterization of regulatory elements controlling expression of the *Bacillus subtilis recA*$^+$ gene. Biochimie 74:755-762.
20. Gutman PD, Carroll JD, Masters CI and Minton KW (1994) Sequencing, targeted mutagenesis and expression of a *recA* gene required for the extreme radioresistance of *Deinococcus radiodurans*. Gene 141:31-37.
21. Fraser CM, Gocayne JD, White O, et al. (1995) The minimal gene complement of *Mycoplasma genitalium*. Science 270:397-403.
22. Stasiak A and Egelman EH (1986) The structure and dynamics of RecA protein-DNA complexes as determined by image analysis of electron micrographs. Biophys J 49:5-7.
23. Stasiak A and Egelman EH (1988) Visualization of recombination reactions. In: Kucherlapati R and Smith GR, eds. Genetic Recombination. Washington, D.C.: American Society for Microbiology, pp 265-307.
24. Dunn K, Chrysogelos S and Griffith J (1982) Electron microscopic visualization of RecA-DNA filaments: evidence for a cyclic extension of duplex DNA. Cell 28:757-765.
25. Egelman EH and Stasiak A (1986) The structure of helical RecA-DNA complexes. I. Complexes formed in the presence of ATP-γ-S or ATP. J Mol Biol 191:677-697.
26. Story RM, Weber IT and Steitz TA (1992) The structure of the *E. coli* RecA protein monomer and polymer. Nature 355:318-325.
27. Koller T, Di Capua E and Stasiak A (1983) Complexes of RecA with single-stranded DNA. In: Cozzarelli N, ed. Mechanisms of DNA replication and recombination. New York: Alan R. Liss, pp 723-729.
28. Shaner SL and Radding CM (1987) Translocation of *Escherichia coli* RecA protein from a single-stranded tail to contiguous duplex DNA. J Biol Chem 262:9211-9219.
29. Lindsley JE and Cox MM (1989) Dissociation pathway for RecA nucleoprotein filaments formed on linear duplex DNA. J Mol Biol 205:695-711.
30. Di Capua E, Engel A, Stasiak A and Koller T (1982) Characterization of complexes between RecA protein and duplex DNA by electron microscopy. J Mol Biol 157:87-103.
31. Dombroski DF, Scraba DG, Bradley RD and Morgan AR (1983) Studies of the interaction of RecA protein with DNA. Nucl Acids Res 11:7487-7504.
32. Stasiak A, DiCapua E and Koller T (1981) The elongation of duplex DNA by RecA protein. J Mol Biol 151:557-564.
33. Register JC III and Griffith J (1985) The direction of RecA protein assembly onto single-strand DNA is the same as the direction of strand assimilation during strand exchange. J Biol Chem 260:12308-12312.
34. Knight KL and McEntee K (1985) Covalent modification of the RecA protein from *Escherichia coli* with the photoaffinity label 8-azidoadenosine 5'-triphosphate. J Biol Chem 260:867-872.

35. Brenner SL, Mitchell RS, Morrical SW, Neuendorf SK, Schutte BC and Cox MM (1987) RecA protein-promoted ATP hydrolysis occurs throughout RecA nucleoprotein filaments. J Biol Chem 262:4011-4016.
36. Pugh BF and Cox MM (1987) Stable binding of RecA protein to duplex DNA. Unraveling a paradox. J Biol Chem 262:1326-1336.
37. Kowalczykowski SC (1991) Biochemistry of genetic recombination: energetics and mechanism of DNA strand exchange. Annu Rev Biophys Chem 20:539-575.
38. Roca AI and Cox MM (1990) The RecA protein: structure and function. Crit Rev Biochem Mol Biol 25:415-456.
39. Little JW, Edmiston SH, Pacelli LZ and Mount DW (1980) Cleavage of the *Escherichia coli* LexA protein by the RecA protease. Proc Natl Acad Sci USA 77:3225-3229.
40. Horii T, Ogawa T, Nakatani T, Hase T, Matsubara H and Ogawa H (1981) Regulation of SOS functions: purification of *E. coli* LexA protein and determination of its specific site cleavage by the RecA protein. Cell 27:515-522.
41. Roberts JW, Roberts CW and Mount DW (1977) Inactivation and proteolytic cleavage of phage λ repressor in vitro in ATP-dependent reaction. Proc Natl Acad Sci USA 74:2283-2287.
42. Sauer RT, Ross MJ and Ptashne M (1982) Cleavage of the lambda and P22 repressors by RecA protein. J Biol Chem 257:4458-4462.
43. Eguchi Y, Ogawa T and Ogawa H (1988) Cleavage of bacteriophage φ80 cI repressor by RecA protein. J Mol Biol 202:565-573.
44. Burckhardt SE, Woodgate R, Scheuermann RH and Echols H (1988) UmuD mutagenesis protein of *Escherichia coli*: overproduction, purification, and cleavage by RecA. Proc Natl Acad Sci USA 85:1811-1815.
45. Little JW (1984) Autodigestion of LexA and phage lambda repressors. Biochemistry 81:1375-1379.
46. Takahashi M and Schnarr M (1989) Investigation of RecA-polynucleotide interactions from the measurement of LexA repressor cleavage kinetics. Eur J Biochem 183:617-622.
47. Lavery PE and Kowalczykowski SC (1988) Biochemical basis of the temperature-inducible constitutive protease activity of the RecA441 protein of *Escherichia coli.* J Mol Biol 203:861-874.
48. Menetski JP, Bear DG and Kowalczykowski SC (1990) Stable DNA heteroduplex formation catalyzed by the *Escherichia coli* RecA protein in the absence of ATP hydrolysis. Proc Natl Acad Sci USA 87:21-25.
49. Silver MS and Fersht AR (1982) Direct observation of complexes formed between RecA protein and a fluorescent single-stranded deoxyribonucleic acid derivative. Biochemistry 21:6066-6072.
50. Zlotnick A, Mitchell RS and Brenner SL (1990) RecA protein filaments bind two molecules of single-stranded DNA with off rates regulated by nucleotide cofactor. J Biol Chem 265:17050-17054.
51. Menetski JP and Kowalczykowski SC (1989) Enhancement of *Escherichia coli* RecA protein enzymatic function by dATP. Biochemistry 28: 5871-5881.

52. Lauder SD and Kowalczykowski SC (1991) Asymmetry in the RecA protein-DNA filament. J Biol Chem 266:5450-5458.
53. Menetski JP and Kowalczykowski SC (1987) Transfer of RecA protein from one polynucleotide to another. Effect of ATP and determination of the processivity of ATP hydrolysis during transfer. J Biol Chem 262:2093-2100.
54. Tsang SS, Chow SA and Radding CM (1985) Networks of DNA and RecA protein are intermediates in homologous pairing. Biochemistry 24:3226-3232.
55. Hsieh P, Camerini-Otero CS and Camerini-Otero RD (1992) The synapsis event in the homologous pairing of DNAs: RecA recognizes and pairs less than one helical repeat of DNA. Proc Natl Acad Sci USA 89:6492-6496.
56. Norioka N, Hsu M-Y, Inouye S and Inouye M (1995) Two *recA* genes in *Myxococcus xanthus*. J Bacteriol 177:4179-4182.
57. Cox MM and Lehman IR (1982) RecA protein-promoted DNA strand exchange. J Biol Chem 257:8523-8532.
58. Cox MM and Lehman IR (1981) Directionality and polarity in RecA protein-promoted branch migration. Proc Natl Acad Sci USA 78: 6018-6022.
59. Kahn R, Cunningham RP, DasGupta C and Radding CM (1981) Polarity of heteroduplex formation promoted by *Escherichia coli* RecA protein. Proc Natl Acad Sci USA 78:4786-4790.
60. West SC, Cassuto E and Howard-Flanders P (1981) Heteroduplex formation by RecA protein: polarity of strand exchanges. Proc Natl Acad Sci USA 78:6149-6153.
61. Das Gupta C and Radding CM (1982) Polar branch migration promoted by RecA protein: effect of mismatched base pairs. Proc Natl Acad Sci USA 79:762-766.
62. Livneh Z and Lehman IR (1982) Recombinational bypass of pyrimidine dimers promoted by the RecA protein of *Escherichia coli*. Proc Natl Acad Sci USA 79:3171-3175.
63. Bianchi ME and Radding CM (1983) Insertions, deletions and mismatches in heteroduplex DNA made by RecA protein. Cell 35:511-520.
64. Jain SK, Cox MM and Inman RB (1994) On the role of ATP hydrolysis in RecA protein-mediated DNA strand exchange. III. Unidirectional branch migration and extensive hybrid DNA formation. J Biol Chem 269: 20653-20661.
65. Kim J-I, Cox MM and Inman RB (1992) On the role of ATP hydrolysis in RecA protein-mediated DNA strand exchange. II. Four-strand exchanges. J Biol Chem 267: 16444-16449.
66. Rosselli W and Stasiak A (1991) The ATPase activity of RecA is needed to push the DNA strand exchange through heterologous regions. EMBO J 10:4391-4396.
67. Kim J-I, Cox MM and Inman RB (1992) On the role of ATP hydrolysis in RecA protein-mediated DNA strand exchange. I. Bypassing a short heterologous insert in one DNA substrate. J Biol Chem 267:16438-16443.

68. Muniyappa K, Shaner SL, Tsang SS and Radding CM (1984) Mechanism of the concerted action of recA protein and helix-destabilizing proteins in homologous recombination. Proc Natl Acad Sci USA 81: 2757-2761.
69. Muniyappa K, Williams K, Chase JW and Radding CM (1990) Active nucleoprotein filaments of single-stranded binding protein and RecA protein on single-stranded DNA have a regular repeating structure. Nucleic Acids Res 18:3967-3973.
70. Lavery PE and Kowalczykowski SC (1992) A postsynaptic role for single-stranded DNA-binding protein in RecA protein-promoted DNA strand exchange. J Biol Chem 267:9315-9320.
71. Hill CW, Harvey S and Gray JA (1990) Recombination between rRNA genes in *Escherichia coli* and *Salmonella typhimurium*. In: Drlica K and Riley M, eds. The Bacterial Chromosome. Washigton, DC: American Society for Microbiology, pp 335-340.
72. Blattner FR, Burland V, Plunkett III G, Sofia HJ and Daniels DL (1993) Analysis of the *Escherichia coli* genome. IV. DNA sequence of the region from 89.2 to 92.8. Nucleic Acids Res 21:5408-5417.
73. Blaisdell BE, Rudd KE, Matin A and Karlin S (1993) Significant dispersed recurrent DNA sequences in the *Escherichia coli* genome. J Mol Biol 229:833-848.
74. Modrich P (1991) Mechanisms and biological effects of mismatch repair. Annu Rev Genet 25:229-253.
75. Feinstein SI and Low KB (1986) Hyper-recombining recipient strains in bacterial conjugation. Genetics 113:13-33.
76. Rayssiguier C, Thaler DS and Radman M (1989) The barrier to recombination between *Escherichia coli* and *Salmonella typhimurium* is disrupted in mismatch-repair mutants. Nature 342:396-401.
77. Shen P and Huang HV (1989) Effect of base pair mismatches on recombination via the RecBCD pathway. Mol Gen Genet 218:358-360.
78. Petit M-A, Dimpfl J, Radman M and Echols H (1991) Control of large chromosomal duplications in *Escherichia coli* by the mismatch repair system. Genetics 129:327-332.
79. Worth L Jr., Clark S, Radman M and Modrich P (1994) Mismatch repair proteins MutS and MutL inhibit RecA-catalyzed strand transfer between diverged DNAs. Proc Natl Acad Sci USA 91:3238-3241.
80. Kodadek T, Wong ML and Alberts BM (1988) The mechanism of homologous DNA strand exchange catalyzed by the bacteriophage T4 UvsX and gene 32 proteins. J Biol Chem 263:9427-9436.
81. Kodadek T and Alberts BM (1987) Stimulation of protein-directed strand exchange by a DNA helicase. Nature 326:312-314.
82. Kodadek T (1991) Inhibition of protein-mediated homologous pairing by a DNA helicase. J Biol Chem 266:9712-9718.
83. Klinkert MQ, Klein A and Abdel-Monem M (1980) Studies on the function of DNA helicase I and DNA helicase II of *Escherichia coli*. J Biol Chem 255:9746-9752.

84. Runyon GT, Bear DG and Lohman TM (1990) *Escherichia coli* helicase II (UvrD) protein initiates DNA unwinding at nicks and blunt ends. Proc Natl Acad Sci USA 87:6383-6387.
85. Zieg J, Maples VF and Kushner SR (1978) Recombination levels of *Escherichia coli* K-12 mutants deficient in various replication, recombination, or repair genes. J Bacteriol 134:958-966.
86. Arthur HM and Lloyd RG (1980) Hyper-recombination in *uvrD* mutants of *Escherichia coli* K-12. Mol Gen Genet 180:185-191.
87. Morel P, Hejna JA, Ehrlich SD and Cassuto E (1993) Antipairing and strand transferase activities of *E. coli* helicase II (UvrD). Nucleic Acids Res 21:3205-3209.
88. Mosig G (1994) Homologous recombination. In: Karam JD, ed. Molecular Biology of Bacteriophage T4. Washington, D.C.: American Society for Microbiology, pp 54-82.
89. Griffith J and Formosa T (1985) The UvsX protein of bacteriophage T4 arranges single-stranded and double-stranded DNA into similar helical nucleoprotein filaments. J Biol Chem 260:4484-4491.
90. Yonesaki T and Minagawa T (1985) T4 phage gene *uvsX* product catalyzes homologous DNA pairing. EMBO J 4:3321-3327.
91. Formosa T and Alberts BM (1986) Purification and characterization of the T4 bacteriophage UvsX protein. J Biol Chem 261:6107-6118.
92. Hinton DM and Nossal NG (1986) Cloning of the bacteriophage T4 *uvsX* gene and purification and characterization of the T4 UvsX recombination protein. J Biol Chem 261:5663-5673.
93. Harm W (1964) On the control of UV-sensitivity of phage T4 by the gene *x*. Mutat Res 1:344-354.
94. Luria SE (1947) Reactivation of irradiated bacteriophage by transfer of self-reproducing units. Proc Natl Acad Sci USA 33:253-264.
95. Kodadek T (1990) The role of bacteriophage T4 gene 32 protein in homologous pairing. J Biol Chem 265:20966-20969.

CHAPTER 3

THE SOS RESPONSE

3.1. THE SOS RESPONSE: THE STRATEGY AND THE TACTICS

How should a cell replicating its chromosome react to the finding that the chromosomal DNA is damaged? Eukaryotic cells choose a prudent approach to this problem, inhibiting DNA replication via a cell-cycle response to ensure sufficient time for one-strand repair (7.0). In contrast, in their race against time, prokaryotes choose an alternative strategy, risking individual survival in exchange for higher rates of DNA replication. The response of prokaryotic cells to DNA damage, which can be summarized as "repair and damage tolerance instead of cell-cycle inhibition", is called the SOS response.

3.1.1. Repair Instead of DNA Damage Checkpoints: The Prokaryotic Strategy

The bulk of two-strand DNA lesions in prokaryotes is generated as a result of DNA replication on template DNA containing one-strand lesions (1.2.1 and 1.2.3). An economical way to prevent the aggravation of DNA damage would be to stop DNA synthesis altogether when one-strand lesions are sensed. This is exactly what eukaryotic cells seem to do — they employ checkpoint mechanisms to delay chromosomal replication when their DNA is damaged (reviewed in refs. 1,2). Since prokaryotes would also benefit from such a strategy, it was argued that *E. coli* might have a system to delay DNA synthesis when its chromosome is damaged.[3]

However, several observations contradict this attractive idea. The initial inhibition of DNA synthesis rate is dose-dependent, while it would be more "all-or-nothing" if a checkpoint mechanism operated. In fact, the rate of DNA replication in *E. coli* is still 20-50% of the maximal rate after irradiation with sublethal UV-doses.[4,5] Furthermore, no *E. coli* mutation has been isolated that prevents the inhibition of chromosomal replication by DNA damage (as *rad9* mutations in yeast or p53 mutations in mammalian cells do (7.0)). Also, preventing initiation of DNA synthesis during irradiation and for a couple of hours thereafter significantly improves the survival of *E. coli*, especially of recombinational repair-deficient mutants.[6-10] If cells would restart DNA replication only when a "safe" level of DNA damage was attained as a result of repair, there would have been no effect of artificial inhibition of DNA synthesis on cell survival. Finally, the idea that *E. coli* has a mechanism to inhibit replication of damaged DNA is incompatible with the observations that *E. coli* initiates extra rounds of DNA replication from the origin when its DNA is heavily damaged.[4,11,12] All these phenomena seem to indicate that, in contrast to eukaryotes, *E. coli* lacks a mechanism to stop chromosomal replication when its DNA is damaged and, instead, relies on enhanced repair and damage tolerance in its attempt to faithfully replicate the damaged genome.

3.1.2. Organization of the SOS Regulon

DNA lesions inhibit DNA replication. In contrast to eukaryotes, where inhibition of DNA replication by lesions is mostly indirect, *E. coli* replisomes seem to be stalled directly at the lesions in template DNA. Inhibition of DNA replication induces the SOS response: an increased expression of some 20 *E. coli* genes aimed at restoring the capacity of the chromosome to replicate (Table 3.1). The resulting enhancement of the cell's ability to repair and tolerate DNA damage is achieved in several independent ways. The cell's capacity for excision repair (1.1.1) is enhanced by overproduction of UvrA and UvrC subunits of UvrABC excinuclease and UvrD helicase. Induced amounts of a minor DNA polymerase, DNA pol II, increases cell's capacity for DNA synthesis across abasic sites.[13,14] The recombinational repair system is beefed up by up to 100-fold increase in the amount of RecA protein (chapter 2) and by a similar increase in the expression of RecN protein.[15] The SOS induction makes possible repeated disengagements of replisomes stalled at the sites of damage in template DNA to allow resumption of synthesis downstream, a phenomenon known as replisome reactivation (4.3). When recombinational repair cannot fix certain DNA lesions, UmuD'C complex catalyzes translesion DNA synthesis (4.6). Overproduction of SfiA protein inhibits cell division[16] providing additional time for completion of recombinational repair. If all these measures fail to overcome inhibition of DNA replication, the lingering SOS induction awakens dormant prophages and prompts them to use a cell with the lethally damaged genome for their lytic development.

Under the normal growth conditions, elevated expression of the SOS genes is prevented by LexA repressor. LexA dimer binds to a palindromic sequence, SOS box, in the promoter regions of the SOS genes, precluding initiation of transcription. The SOS box has an inverted repeat consensus 5' — **TACTG**TATATATATA**CAGTA** — 3', where the positions in bold are absolutely conserved.[17] The substantial uninduced levels of certain SOS gene products (Table 3.1) are maintained due to imperfect SOS boxes in their operator regions. Tight regulation of the genes with low-affinity SOS boxes is achieved by the high intracellular concentration of LexA repressor (more than 1000 molecules per cell)[18] and by the presence of two SOS boxes in the operators of *lexA* and *recN*.[19]

The increased expression of the SOS genes in response to DNA synthesis inhibition is a result of LexA repressor inactivation. The inactivation of LexA repressor is by autocleavage catalyzed by a specific form of RecA protein (2.2.2). This specific form of RecA is thought to be the extended RecA filament assembled on ssDNA (2.2.1). Only LexA molecules that are free in solution can be inactivated,[20] which accounts for the late induction of the SOS genes with high-affinity SOS-boxes.

3.1.3. The Levels of SOS Induction

The SOS response is by no means a desperate attempt of a cell to stay alive, but rather an orderly and measured reaction to inhibition of DNA synthesis. General information on the *E. coli* genes induced during the SOS response is summarized in Table 3.1. The strength of SOS boxes in the operator regions of SOS genes correlates well with in vitro LexA repressor affinity to the corresponding promoters and is likely to reflect the timing of expression of a given gene during SOS induction. According to the inferred time frames of expression during SOS induction, the genes of the SOS regulon could be subdivided into three groups. The first genes to be induced are mostly those responsible for one-strand repair (*uvrA*, *uvrB*, *uvrD*) or damage tolerance (*polB*) (Fig. 3.1). Expression of LexA repressor itself is also induced immediately. The levels of enhanced expression of the immediately induced genes are usually less than 10 times of their regular expression. If the increased expression of the one-strand repair genes does not help to regain normal rates of DNA synthesis, the genes of recombinational repair, *recA* and *recN*, are induced (Fig. 3.1). The induction of these genes can be higher — 20- to 50-fold over their regular levels. When DNA damage is massive, so even enhanced recombinational repair needs some time to overcome the inhibition of DNA replication, the third group of genes represented by *sfiA* and *umuDC* is called into action. Since these last activities are maintained at very low levels during normal DNA synthesis, their SOS induction could be more than 100-fold. The expression of *umuDC* operon decreases the level of recombinational repair and makes possible translesion DNA synthesis, while SfiA protein inhibits cell division. When DNA replication rates begin to normalize, the three expression groups of the SOS genes become repressed in the reversed order (Fig. 3.1).

Table 3.1. E. coli *proteins with known functions induced during the SOS response*

Gene	Gene product/Function	Basal level (copies/cell)	SOS-induced level (copies/cell)	Strength of SOS box[a]/LexA affinity[b]	Increase in expression[c]
Expressed first					
lexA	LexA/SOS repressor	1,300	1*	6.4 & 8.3 / 15	5.8
uvrA	Subunit of UvrABC excinuclease/ Excision repair	20	250	7.0 / 14.6	4.8
uvrB	Subunit of UvrABC excinuclease/ Excision repair	250	1,000	6.1 / 8.8	3.7
uvrD	Helicase II/Excision repair, fidelity of rec. repair	5,000-8,000	25,000-65,000	8.8 / 17.9	5.9
polB	DNA polymerase II/ Translesion DNA synthesis	?	7-fold	12.1 / ?	7.3
ruvA	Subunit of RuvAB helicase/ Recombinational repair	?	?	9.2 / ?	2-3
ruvB	Subunit of RuvAB helicase/ Recombinational repair	?	?	see ruvA	
Expressed second					
recA	RecA coprotease, synaptase/ SOS de-repressor, rec. repair	800-1,500	100,000	4.3 / 3.8	12.0
recN	RecN / Recombinational repair	?	?	4.2 & 9.4 / ?	10

Expressed last					
sfiA	SfiA (SulA)/ Cell division inhibitor	?	?	4.7 / 1	125
umuD	Subunit of UmuD'C/ Translesion DNA synthesis	180	2400	2.8 / 1.1	22.5
umuC	Subunit of UmuD'C/ Translesion DNA synthesis	0	200	see umuD	

* LexA protein is degraded during SOS induction

a The strength of the SOS box in the promoter region of the gene is represented by the heterology index. Higher values reflect more deviations from the SOS box consensus, thus weaker LexA binding. The data are from Lewis et al (1994) J Mol Biol 241:507-523.

b LexA affinity is expressed as relative LexA affinity in vitro compared to the affinity to *sfiA* operator. The highest values mean weak LexA binding, while the lowest values mean strong LexA binding. The data are from Schnarr et al (1991) Biochimie 73:423-431.

c Increase in expression is the ratio of the gene expression without LexA repression to the gene expression with full repression. Except for *ruvA* and *recN*, the values are means of the measurements done at 30°C and 42°C (from Schnarr et al (1991) Biochimie 73:423-431).

? not known.

References

LexA: Sassanfar and Roberts (1990) J Mol Biol 212:79-96.
UvrA and UvrB: Van Houten (1990) Microbiol Rev 54:18-51.
UvrD: Klinkert et al (1980) J. Biol. Chem. 255:9746-9752; Kumura and Sekiguchi (1984) J Biol Chem 259:1560-1565.
PolB: Bonner et al (1988) J Biol Chem 263:18946-18952.
RuvAB: Shurvinton and Lloyd (1982) Mol Gen Genet 185:352-355.
RecA: Karu and Belk (1982) Mol. Gen. Genet. 185:275-282; Salles and Paoletti (1983) Proc Natl Acad Sci USA 80:65-69.
RecN: Picksley et al (1984) Cold Spring Harbor Symp Quant Biol 49:469-474.
UmuD and UmuC: Woodgate and Ennis (1991) Mol. Gen. Genet. 229:10-16.

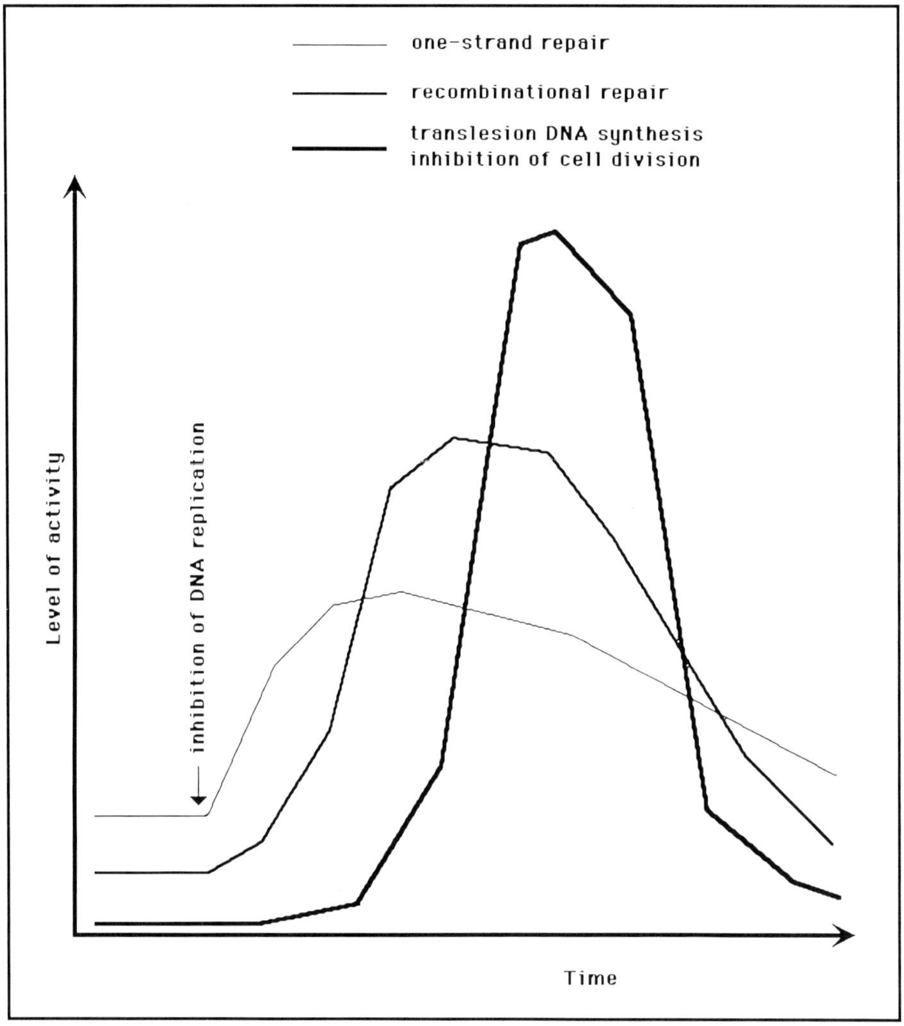

Fig. 3.1. A scheme of induction kinetics of the three groups of SOS genes. This highly idealized picture is intended only to demonstrate the well-regulated nature of the SOS response. Both X-axis (time) and Y-axis (the level of SOS induction) are in arbitrary units, and the heights of the three curves are not comparable.

3.2. A POSSIBLE MECHANISM OF SOS INDUCTION

The exact mechanism of SOS induction is still unclear. It is believed that DNA damage results in the appearance of ssDNA, which allows RecA filament assembly and, hence, LexA cleavage.[18,21,22] However, in certain conditions DNA inhibition alone seems to be sufficient for SOS induction. Therefore, it is suggested that the SOS response is induced when two conditions are met: (1) DNA synthesis is inhibited and (2) single-stranded DNA is available. The system that senses the SOS conditions is likely to be based on competition of SSB and RecA proteins for binding to single-strand DNA (ssDNA).

3.2.1. In Vivo, RecA Binds ssDNA Only when DNA Synthesis Is Inhibited

A growing *E. coli* cell always has some of its DNA in a single-stranded form. This ssDNA is continuously present at several replication forks. In rapidly growing cells there are from 2-6 replication forks (1.3) with the total amount of ssDNA from 2-12 kb. Why does RecA not form a filament on this DNA in the absence of DNA damage?

One answer may be that there is not enough RecA in uninduced conditions to cover this ssDNA. However, the basal level of RecA protein in cells uninduced for SOS is from 1,200-13,000 monomers per cell, as measured by different assays.[18,23-25] Assuming the stoichiometry of ssDNA binding of 3 nucleotides per RecA monomer (2.2.1) this amount of RecA appears to be enough to cover all the regions of ssDNA which are present at the several replication forks. Thus, there should be another reason why in the absence of DNA damage RecA does not polymerize on these ssDNA regions.

This reason may be that ssDNA at replication forks is likely to be protected by single-strand DNA binding protein (SSB). There are 1000-2000 SSB tetramers per *E. coli* cell,[26] which is enough to complex up to 150 kb of ssDNA.[27] Thus, SSB is in a huge excess to ssDNA, and under normal circumstances, RecA may be unable to polymerize on ssDNA because it is outcompeted by SSB.

When bacterial DNA is damaged (X-rays, UV), the ongoing DNA synthesis abruptly stops, while RNA and protein syntheses continue.[28,29] Although the amount of SSB is not increased significantly in response to DNA damage,[26,30,31] there should be enough SSB to complex all the ssDNA generated by a very limited replication through a damaged template DNA. Nevertheless, if DNA is damaged, the SOS response is induced, indicating that RecA now gains access to ssDNA.

A trivial explanation for this observation is that the amount of SSB is not enough to complex all the ssDNA formed. However, there are two additional observations making this explanation unlikely. First one is made with strains overproducing SSB about 30-fold, so there is enough of it to complex half of the entire chromosome if rendered single-stranded. In such strains, DNA damage still induces the SOS response to approximately the same levels and with similar kinetics.[25] The second observation is made with cells in which DNA synthesis is inhibited by DNA cross-linking or by shifting *dnaB*, *dnaC*, *dnaE* or *dnaG* temperature-sensitive mutants to nonpermissive temperatures. In these cells, the amount of ssDNA must not change with time, as all of it is present at the same number of replication forks, which, at least in *dnaE* and *dnaG* mutants, are stable (the newly synthesized

DNA is not degraded).[32] The ssDNA at these stalled replication forks must be still complexed with SSB; nevertheless, the SOS response is induced.[18,32,33] Thus, inhibition of DNA replication seems to tip the balance of RecA-SSB competition over ssDNA in favor of RecA.

3.2.2. Inhibiting DNA Synthesis Causes Accumulation of DNA Precursors and Their Counterions

Inhibition of DNA synthesis should lead to accumulation of DNA precursors, deoxyribonucleotides. Elevated concentrations of DNA precursors, especially dATP, were proposed to tip the balance of SSB-RecA competition, starting the SOS response.[34,35] This explanation is supported by the observations that dATP is a superior to rATP cofactor for all the *E. coli* RecA-promoted reactions tested (ref. 35 and references therein). Indeed, up to 30-fold accumulation of dATP is detected after inhibition of DNA synthesis.[36-40]

Since deoxyribonucleotides (dNTPs) are produced from ribonucleotides (NTPs) by an enzyme which is inhibited by unused dNTPs,[41] DNA synthesis inhibition should also result in accumulation of NTPs. The combined nucleotide pool of the fast-growing cells of the *E. coli* relative, *Salmonella typhimurium*, is measured to be about 15 µmol/g, out of which only about 1 µmol/g is in deoxyribonucleotides.[42] The amount of chromosomal DNA (in bases) is about 100 µmol/g.[43] If the doubling time of a bacterial culture is 30 minutes, cells are incorporating 100 µmol/g of the precursors into DNA every half an hour (we do not take into account phosphates that are cycled in the reaction). Inhibiting DNA synthesis in these conditions even for 5 minutes would create a backlog of dNTP and rNTP equal in size to the regular combined nucleotide pool.

rATP represents roughly a half of the cell's combined nucleotide pool.[39,42] Inhibition of DNA synthesis with UV-irradiation, bleomycin, or nalidixic acid treatment causes a transient 2- to 3-fold increase in the intracellular rATP concentration.[44-46] Similarly, in conditions, nonpermissive for T4 amber mutants in replicative DNA helicase or T4 DNA polymerase, there is an up to 5-fold expansion of the intracellular ribonucleotide pools.[39] Thus, inhibiting DNA synthesis indeed increases the intracellular nucleotide pool.

DNA is a polyanion, and its charges need to be countered by cations and basic proteins. The charges of DNA and its precursors in bacterial cells are countered mostly by Mg^{2+} and polyamines (discussed in ref. 47). The intracellular concentration of soluble Mg^{2+}, which is all derived from the growth medium, is reported to be about 4 mM[48] or 1-2 mM[49,50] (however, see ref. 51). The intracellular concentrations of free polyamines are not known. It was proposed[47] that a block to DNA replication, causing the accumulation of DNA precursors, is likely to lead to a proportional increase in the intracellular concentration of their counterions, polyamines and Mg^{2+}. Consequently, it could be the parallel rise of free polyamine, Mg^{2+} and ATP concentrations that enables RecA to polymerize on ssDNA occupied by SSB.

Indeed, in vitro, RecA can not polymerize on SSB-bound ssDNA in 1 mM Mg^{2+},[52] but can do so if 1 mM spermidine is added to the reaction.[53] Polymerization of RecA on SSB-bound ssDNA is also stimulated in vitro simply by raising Mg^{2+} concentration above 4 mM[52,53] arguing against specific interactions between RecA and spermidine. The high concentration of ATP is also important since in vitro, in the presence of 12 mM Mg^{2+}, RecA gains access to the SSB-bound ssDNA only if ATP concentration is 2 mM or higher with the maximal rate of RecA polymerization at 8 mM ATP being some 30 times higher than at 3 mM ATP.[54] Interestingly, regular intracellular concentration of ATP is about 3 mM,[39,42,55] while the 2.5- to 3-fold increase in ATP concentration observed after DNA synthesis inhibition[39,44-46] should bring it up to about 8 mM.

3.2.3. SSB Binds ssDNA with Two Different Stoichiometries

RecA polymerization on SSB-bound ssDNA during the SOS response could be brought about in two different ways. On one hand, the ability of RecA to polymerize on ssDNA might stay the same, but the binding of SSB to DNA could be modified at higher concentrations of Mg^{2+} and spermidine. Alternatively, SSB binding might be unchanged, but RecA could form a filament more readily at higher ionic strength and hence, ignore the presence of SSB. In vitro experiments support both possibilities.

The mode and cooperativity of SSB binding to ssDNA is sensitive to ionic conditions. This protein is known to bind ssDNA with different site sizes at different concentrations of monovalent and divalent cations. The smallest site size corresponds to SSB tetramer binding about 35 nucleotides; this binding is cooperative[56] and is observed at low ionic strength. The biggest site size is almost twice as big, with 65 nucleotides per SSB tetramer; this mode of binding displays only limited cooperativity and is observed at high ionic strength. Which binding mode is predominant in vivo is not clear since the in vitro transition from SSB_{35} to SSB_{65} mode is a broad one and happens within the range of physiological concentrations of both mono- and divalent cations.[57] Spermidine also induces the transition to the "big site" binding mode, but in concentrations that are 10 times lower than those of Mg^{2+}, which singles it out as a possible major effector in this transition.[58] It is proposed that cooperative, small-size binding mode of SSB is important for DNA replication, while big-size binding mode is used during DNA repair.[59] This proposal is substantiated by the finding that cells carrying a mutant SSB protein which in vitro in physiological conditions exists only in the small size binding mode[60] are viable but severely deficient in repair of UV-damage.[61]

The proposed accumulation of spermidine and Mg^{2+} after DNA synthesis inhibition is likely to shift the equilibrium of SSB binding to ssDNA toward SSB_{65} mode. Electron microscopic observations show that in conditions mimicking physiological ones (saturating amounts of SSB, concentrations of Mg^{2+} lower than 5 mM) SSB completely engulfs ssDNA within a smooth, continuous complex.[62] However, when the concentration of Mg^{2+} is raised above 5 mM, the SSB-ssDNA complex assumes a beaded appearance, with some free DNA between the beads. It seems straightforward to relate smooth complex to the $(SSB)_{35}$ binding mode and the beaded complex — to the $(SSB)_{65}$ binding mode. Maybe coincidentally,[63] electron microscopic observations show that RecA filament can assemble on a ssDNA covered by the beaded SSB complex, but not by the smooth SSB complex.[62] This is further corroborated by in vitro data that SSB cannot displace RecA from ssDNA at Mg^{2+} concentrations higher than 4 mM.[52,64] Thus, changes in the pattern of SSB binding to ssDNA due to increases in concentrations of Mg^{2+} and spermidine are likely to be a part of the SOS induction mechanism.

3.2.4. RecA Attains Its Full Range of In Vitro Activities at Elevated Mg^{2+} Concentrations and in the Presence of Spermidine

Among the properties of RecA which can be monitored in vitro are protein polymerization, DNA binding to chemically modified ssDNA, DNA-dependent ATPase, cleavage of LexA and phage repressors, homology-independent coaggregation of DNA, and homology-dependent strand invasion and strand exchange reactions. All these reactions are strongly dependent on Mg^{2+} (although their magnesium-dependence profiles vary), and some of them are stimulated by additional spermidine.

In vitro, a narrow range of Mg^{2+}, Ca^{2+} or Mn^{2+} concentrations (10-20 mM) or spermidine in concentrations above 3 mM induce extensive RecA polymerization and even precipitation from crude cell lysates as well as from purified preparations.[65-67] This spermidine- or Mg^{2+}-induced RecA polymerization is not inhibited by high concentrations of other proteins.

RecA protein binding to ssDNA and its DNA-dependent ATPase activity are readily observed at 1 mM Mg^{2+}, although both activities are inhibited at lower Mg^{2+} concentrations.[68,69] ATPase activity of RecA on native ssDNA is not stimulated by Mg^{2+} concentration higher than 1 mM[68,70] because of the increasing formation of secondary structures in ssDNA at higher Mg^{2+} concentrations which disrupt RecA filaments.[71] Formation of secondary structures is impossible in chemically-modified ssDNA called etheno-DNA. Interestingly, ATPase activity of RecA on etheno-DNA is stimulated 2-2.5 times by increasing Mg^{2+} concentration from 1 mM to 4 mM.[70] Moreover, the rate of RecA binding to etheno-DNA increases linearly with increasing magnesium ion concentration, being an order of magnitude higher at 8 mM Mg^{2+} than at 1 mM.[72] RecA-promoted cleavage of LexA repressor in vitro is usually carried out at Mg^{2+} concentrations between 4 and 10 mM,[73,74] although LexA cleavage by RecA at 2 mM Mg^{2+} is reported.[75] The other RecA activities: the cleavage of phage repressors, DNA coaggregation, and strand invasion and exchange, are only observed at Mg^{2+} concentrations higher than 4 mM, with optimal concentrations around 10 mM.[76-79] Thus, it seems that under "normal" conditions, the only reaction RecA can perform is the cleavage of LexA repressor. In all other aspects, RecA appears to be fully active only at the elevated Mg^{2+} concentrations and in the presence of spermidine, i.e., in conditions when RecA can polymerize on SSB-covered ssDNA.

3.3. Structural Transitions within RecA Monomer and Regulation of RecA Polymerization by Other Proteins

Above it was suggested that inhibition of DNA synthesis leads to accumulation of RNA and DNA precursors and their counterions, Mg^{2+} and spermidine. Simultaneous rise in the concentration of ATP, Mg^{2+} and spermidine is likely to change the in vivo modes of both SSB and RecA binding to ssDNA, as it does in vitro. Under normal conditions, RecA can not form stable polymers on ssDNA occupied by SSB (although will probably polymerize on SSB-free DNA). We propose that the ATP-, Mg^{2+}- and spermidine-induced changes allow RecA to form filaments on SSB-bound ssDNA and to induce the SOS response (Table 3.2).

The regulation of RecA polymerization by Mg^{2+} concentration in vitro works both ways: if RecA is assembled on ssDNA in the absence of SSB at a low Mg^{2+} concentration, subsequent addition of SSB leads to a displacement of RecA from ssDNA.[52] Thus, the ability of RecA to form a stable filament on ssDNA in the presence of SSB seems to be finely tuned to the concentrations of specific ions. What could be the basis for this ability of RecA for reversible polymerization?

3.3.1. The RecA Potential for Filament Assembly Could be Modulated by the Protein's C-Terminus

Since the full range of RecA activities is observed only at high Mg^{2+} concentrations, stability of RecA filament is apparently improved under these conditions. According to RecA crystal structure,[80] a RecA monomer can be diagrammed as a two-lobe globule (Fig. 3.2). It has two surfaces that are responsible for the monomer-monomer interactions during RecA filament assembly. The interactions between monomers in a RecA polymer are mostly hydrophobic.[81,82] However, the crystal structure of RecA protein reveals that the surfaces for monomer-monomer interaction, though mostly hydrophobic, also have charged amino acids.[80] In fact, one surface is negatively charged (9 acidic residues, 2 basic

Table 3.2. In vitro properties of SSB and RecA proteins of* Escherichia coli *and their suggested interplay in the SOS induction

DNA synthesis	Normal	Inhibited
	↓	↓
Intracellular ATP concentration	low ↓	high ↓
Intracellular spermidine and Mg^{2+} concentration (suggested)	low ↓	high ↓
Mode of SSB binding to ssDNA	SSB_{35}	SSB_{65}
Binding of RecA to ssDNA*	slow	fast
ATPase activity of RecA on ssDNA*	low	high
RecA polymerization on ssDNA in the presence of SSB	-	+
Cleavage of phage repressors by activated RecA	-	+
RecA-promoted strand invasion	-	+
SOS induction	-	+

* etheno-DNA
Adapted with permission from J Theor Biol (1995) 177: 29.

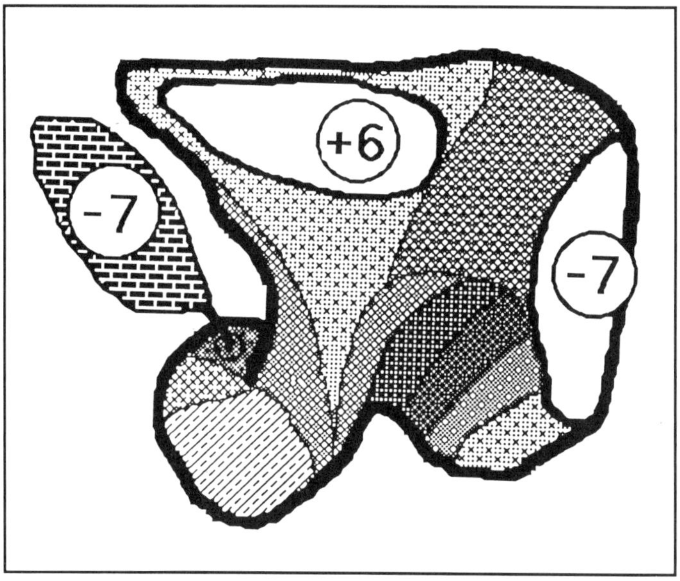

Fig. 3.2. A schematic representation of RecA monomer, as adapted from RecA crystal structure.[80] The open areas with the indicated charges are the two hydrophobic surfaces participating in the polymer assembly. The negatively charged C-terminal domain is depicted as a leaf near the positively charged surface. The differential shading is used to imitate the 3D structure of RecA monomer. Reprinted with permission from J Theor Biol (1995) 177: 29.

residues), while the other one is charged positively (6 basic residues, no acidic residues). This makes RecA monomer a dipole and allows only head-to-tail interactions between monomers in the polymer.

Another interesting feature of the RecA monomer is that it has a disordered 25 aminoacid C-terminal domain[80] containing no basic residues but 7 acidic residues. Since this negatively-charged flexible tail is in direct proximity to the positively charged hydrophobic surface for the monomer-monomer stacking (Fig. 3.2), it is suggested to interfere with RecA polymerization by masking this surface[47] (Fig. 3.3). Under "normal" in vivo conditions RecA might be unable to polymerize on SSB-bound ssDNA because it is only partly active for assembly, although it apparently can polymerize on naked ssDNA.[83,84] However, inhibiting DNA replication could increase the intracellular concentration of Mg^{2+} and polyamines, which would shield the negative charges of the RecA C-terminal domain (Fig. 3.3). This should stimulate RecA polymerization and tip the balance of SSB-RecA competition for ssDNA in favor of RecA.

3.3.2. A Way for the Regulation of RecA Polymerization by the Acidic Domains of RecA and SSB

The sequence alignment of 23 known bacterial RecA proteins shows that the C-terminal domains, presented in 20 proteins, show very little sequence similarity.[85] This makes it even more significant that all the C-terminal domains of prokaryotic RecA proteins are devoid of basic residues and are loaded with acidic residues (from 4-11 of them in a 20 or less aminoacid stretch). Thus, the negatively-charged tail is not peculiar to the RecA protein of *E. coli.*

The acidic C-terminal domain of RecA could negatively affect RecA polymerization by competing for the positively charged hydrophobic sites for monomer-monomer interaction in a polymer. By doing so, it would discourage RecA polymerization in suboptimal conditions in two ways. First, it would slow down RecA polymerization by masking the positively-charged site for monomer-monomer interaction, and second, it would destabilize the existing RecA polymers (Fig. 3.3).

Although the crystal structure of SSB protein is not yet available, it is believed that the protein consists of a bigger N-terminal domain, which is compact and is responsible for the DNA-binding and tetramer formation, and a smaller C-terminal domain, which is less structured, negatively charged, and is thought to regulate SSB interactions with other proteins.[26] The N-terminal domain alone, if separated from the C-terminal domain by proteolysis, can bind ssDNA with an efficiency, equal to or greater than that of the whole protein.[86] In vitro, the truncated SSB protein is superior to WT SSB in stimulating RecA-promoted reactions.[87] The negatively-charged C-terminal domain of SSB could also inhibit RecA polymerization on ssDNA by competing for the binding to the positively-charged RecA surface for monomer-monomer interaction.

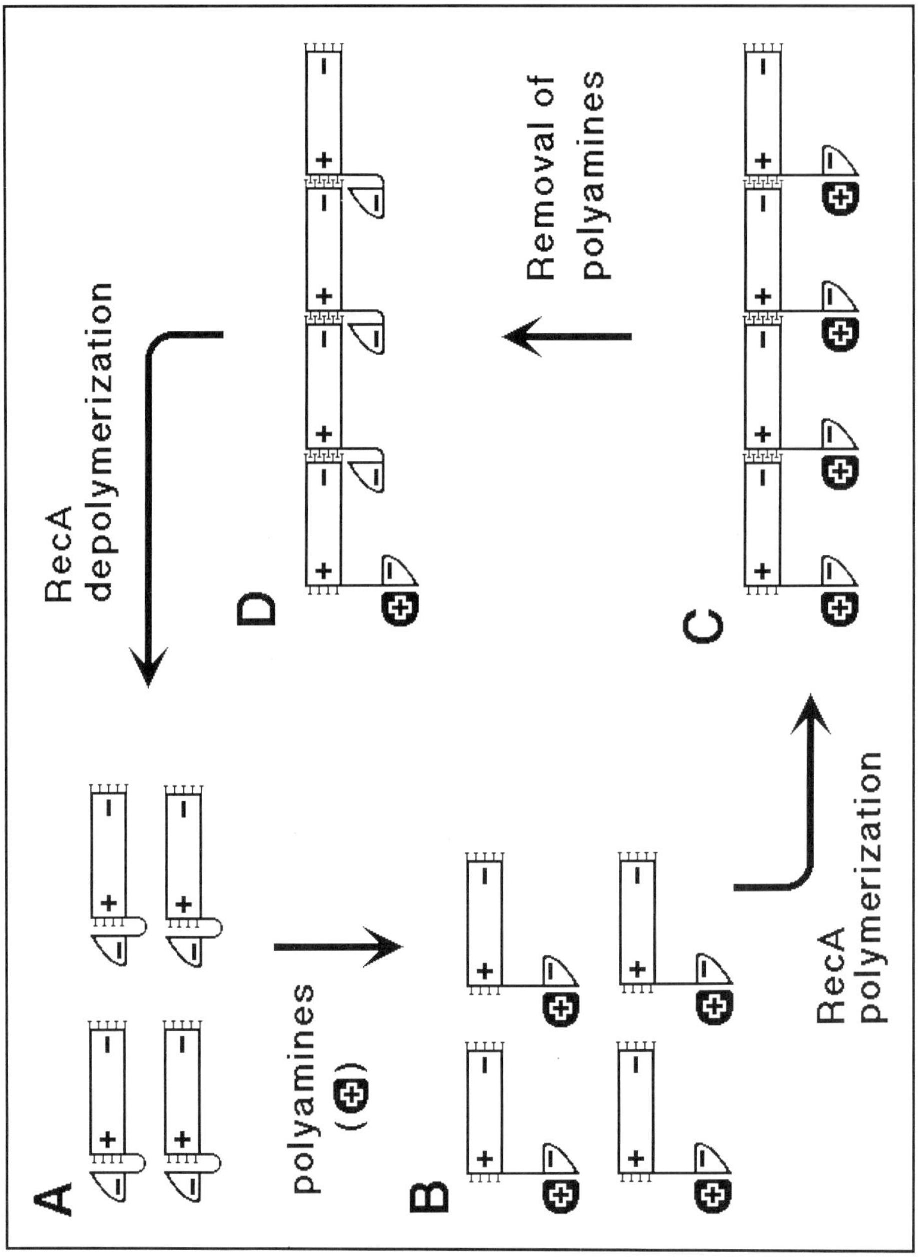

Fig. 3.3. Regulation of RecA polymerization by Mg^{2+} and polyamines. (A) RecA monomers can not form a polymer since their positively charged hydrophobic surfaces for polymerization are capped with the negatively charged C-terminal domains. (B) Polyamines and Mg^{2+} neutralize the negatively-charged caps, converting RecA monomers into dipoles. (C) RecA monomers polymerize using their charged hydrophobic surfaces to determine the direction of polymerization. (D) When the concentration of Mg^{2+} and polyamines falls back to normal, the C-terminal domains sever the RecA filament. Reprinted with permission from J Theor Biol (1995) 177:29.

3.3.3. Regulation of RecA Polymerization by Other Proteins

The proposed autoregulation of RecA filament assembly/disassembly is potentially susceptible to the influence of other proteins. For example, a protein with a positively-charged domain would be able to promote RecA polymerization on SSB-covered ssDNA. Such a protein, if real, should act early in the RecA-promoted recombinational repair reaction. It could be used to organize RecA filament on ssDNA independently of whether the situation calls for it or not. If DNA synthesis is not inhibited, the nascent RecA filament will be unstable (3.3.2), so there will be no consequence for the cell. However, if DNA synthesis is inhibited, the nascent RecA filament will be stabilized and will be capable of catalyzing LexA cleavage and strand exchange (Fig. 3.4). The amount of such a protein assisting in RecA filament assembly does not need to be increased during the SOS response, since if anything, RecA polymerization is probably facilitated in the SOS-inducing conditions.

Following the same logic, a protein with a negatively-charged domain could be helpful in promoting RecA filament disassembly. Such a protein, if exists, must act late in recombinational repair, removing used RecA filaments from DNA. Since the need for a RecA-dispersing protein would be higher in conditions that encourage RecA binding to DNA (like those during SOS induction), the expression of such a protein must be induced during the SOS response. As we will see in the next two chapters, several accessory proteins in the two main pathways of recombinational repair are indeed capable of catalyzing RecA filament assembly or disassembly.

SUMMARY

The *E. coli* strategy for dealing with replication of damaged DNA is to wait until the replisomes are blocked at DNA lesions. Then the cells would induce the SOS response — an increased expression of some 20 genes, aimed at boosting cell's capacity for DNA repair and damage tolerance. The increased expression of the genes of the SOS regulon is a result of inactivation of the LexA repressor, which binds to SOS boxes in promoter regions of these genes. LexA inactivation is due to LexA autocleavage promoted by RecA filaments polymerized on ssDNA. The SOS response is an orderly and balanced reaction, which develops in three phases. A weak inhibition of DNA synthesis leads mostly to the enhancement of one-strand repair. A substantial inhibition of DNA synthesis results in an additional enhancement of recombinational repair. A prolonged inhibition of DNA synthesis prompts the expression of translesion DNA synthesis and inhibits cell division. The main elements of the SOS response are likely to be conserved among bacteria, as illustrated by the striking sequence homology among the SOS boxes of gram-negative bacteria[85] and by the characterization of SOS boxes of a different sequence in the operator regions of damage-inducible *Bacillus subtilis* genes.[88]

The signal for the SOS response in bacteria has always been a mystery.[22] While early speculations entertained the possibility that the SOS inducing signal is an accumulation of DNA precursors as a result

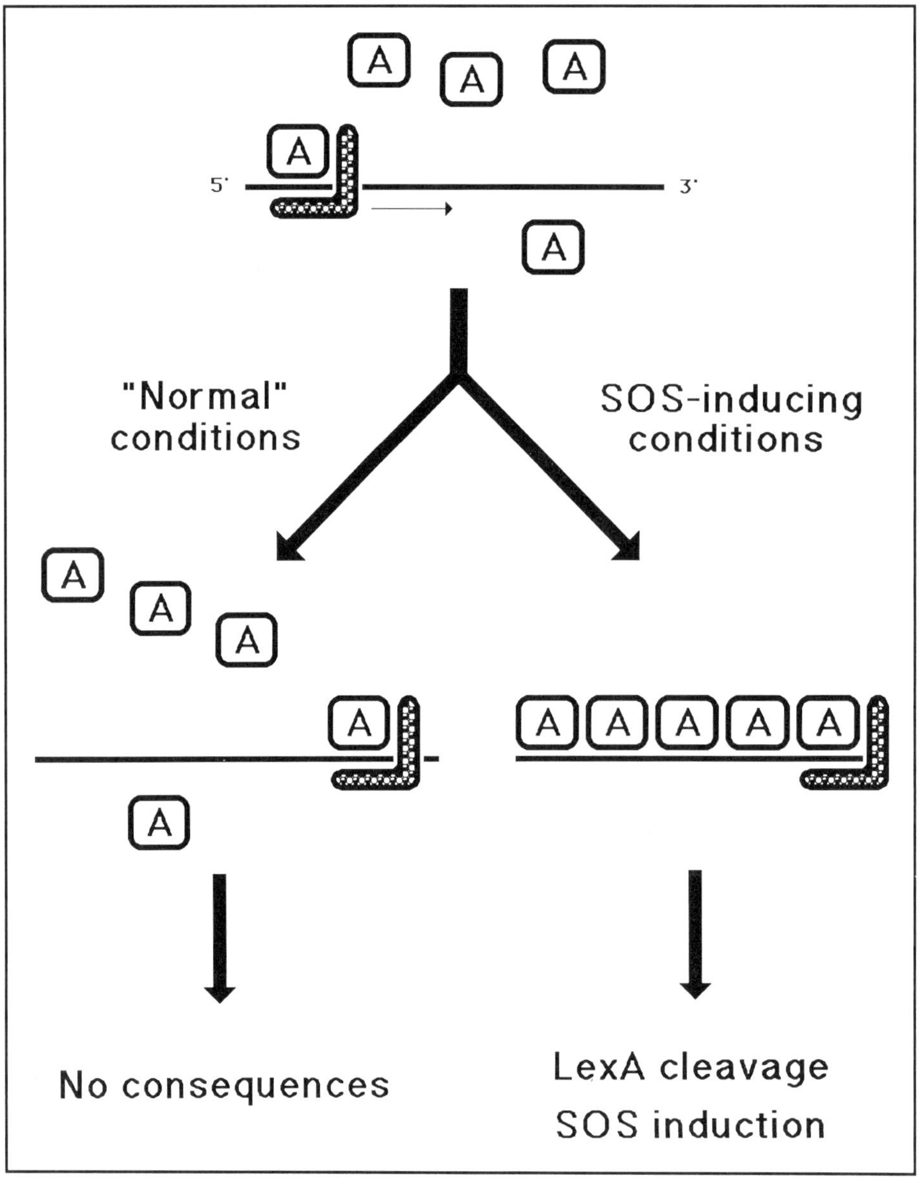

Fig. 3.4. A plausible mechanism for SOS induction. Open rectangles with a letter "A" – RecA monomers; a boomerang moving along the DNA – a protein assisting in RecA polymerization. Explanations in the figure and in the text.

of inhibition of DNA synthesis,[89] subsequent ideas were influenced by the in vitro properties of RecA, especially its preference for ssDNA. Thus, the last conclusion was that the SOS response is induced by an increased amount of single-stranded DNA, formed as a result of DNA damage processing.[18,90] However, while this explanation might work when DNA is indeed damaged and degraded, it does not explain the induction of the SOS response by a simple block to DNA synthesis in conditions where no DNA degradation is detected.[18,32,33]

A plausible mechanism for SOS induction is outlined above (Fig. 3.4). RecA protein, assisted by other proteins, always tries to polymerize on ssDNA, which is present at replication forks. When DNA synthesis is not inhibited, nascent RecA filaments are not stable and can not cause appreciable LexA cleavage. When DNA synthesis is inhibited, accumulation of DNA precursors and their counterions stabilizes nascent RecA filaments, resulting in LexA cleavage and SOS induction (Fig. 3.4). When DNA lesions inhibiting DNA replication are repaired, the composition of the intracellular sap returns back to normal rendering the nascent RecA filaments unstable.

The negatively charged C-terminal domain of RecA is proposed to participate in the regulation of RecA polymerization in vivo by modulating the polymerization potential of RecA monomers. A corollary to this proposal is that there could be accessory recombinational repair proteins whose main function is to promote RecA polymerization or depolymerization. The next chapter amply illustrates the important role of RecA-assisting proteins in recombinational repair of daughter-strand gaps.

References

1. Kastan MB and Kuerbitz SJ (1993) Control of G1 arrest after DNA damage. Environ Health Perspect 101, Suppl. 5:55-58.
2. Maity A, McKenna WG and Muschel RJ (1994) The molecular basis for cell cycle delays following ionizing radiation: a review. Radiother Oncol 31:1-13.
3. Bridges BA (1995) Are there DNA damage checkpoints in *E. coli*? BioEssays 17:63-70.
4. Khidhir MA, Casaregola S and Holland IB (1985) Mechanism of transient inhibition of DNA synthesis in ultraviolet-irradiated *E. coli*: inhibition is independent of *recA* whilst recovery requires RecA protein itself and an additional, inducible SOS function. Mol Gen Genet 199:133-140.
5. Doudney CO (1990) DNA-replication recovery inhibition and subsequent reinitiation in UV-radiation-damaged *E. coli*: a strategy for survival. Mutat Res 243:179-186.
6. Billen D (1963) Unbalanced deoxyribonucleic acid synthesis: its role in X-ray-induced bacterial death. Biochim Biophys Acta 72:608-618.
7. Ginsberg DM and Jagger J (1965) Evidence that initial ultraviolet lethal damage in *Escherichia coli* strain 15 $T^-A^-U^-$ is independent of growth phase. J Gen Microbiol 40:171-184.

8. Hanawalt PC (1966) The U.V. sensitivity of bacteria: its relation to the DNA replication cycle. Photochem Photobiol 5:1-12.
9. Pollard EC, Bronner CE and Fluke DJ (1986) The influence of repair systems on the presence of sensitive and resistant fractions in the relation of survival of colony-forming ability in *Escherichia coli* to UV exposure. Mutat Res 165:63-70.
10. Smith KC and Sharma RC (1987) A model for the *recA*-dependent repair of excision gaps in UV-irradiated *Escherichia coli*. Mutat Res 183:1-9.
11. Pritchard RH and Lark KG (1964) Induction of replication by thymine starvation at the chromosome origin in *Escherichia coli*. J Mol Biol 9:288-307.
12. Billen D (1969) Replication of the bacterial chromosome: location of new initiation sites after irradiation. J Bacteriol 97:1169-1175.
13. Bonner CA, Randall SK, Rayssiguier C, Radman M, Eritja R, Kaplan BE, McEntee K and Goodman MF (1988) Purification and characterization of an inducible *Escherichia coli* DNA polymerase capable of insertion and bypass at abasic lesions in DNA. J Biol Chem 263:18946-18952.
14. Tessman I and Kennedy MA (1994) DNA polymerase II of *Escherichia coli* in the bypass of abasic sites in vivo. Genetics 136:439-448.
15. Picksley SM, Lloyd RG and Buckman C (1984) Genetic analysis and regulation of inducible recombination in *Escherichia coli* K-12. Cold Spring Harbor Symp Quant Biol 49:469-474.
16. Bi E and Lutkenhaus J (1993) Cell division inhibitors SulA and MinCD prevent formation of the FtsZ ring. J Bacteriol 175:1118-1125.
17. Lewis LK, Harlow GR, Gregg-Jolly LA and Mount DW (1994) Identification of high affinity binding sites for LexA which define new DNA damage-inducible genes in *Escherichia coli*. J Mol Biol 241:507-523.
18. Sassanfar M and Roberts JW (1990) Nature of the SOS-inducing signal in *Escherichia coli*. The involvement of DNA replication. J Mol Biol 212:79-96.
19. Schnarr M, Oertel-Buchheit P, Kazmaier M and Granger-Schnarr M (1991) DNA binding properties of the LexA repressor. Biochimie 73:423-431.
20. Little JW and Mount DW (1982) The SOS regulatory system of *Escherichia coli*. Cell 29:11-22.
21. Walker GC (1984) Mutagenesis and inducible responses to deoxyribonucleic acid damage in *Escherichia coli*. Microbiol Rev 48:60-93.
22. Peterson KR, Ossanna N, Thliveris AT, Ennis DG and Mount DW (1988) Derepression of specific genes promotes DNA repair and mutagenesis in *Escherichia coli*. J Bacteriol 170:1-4.
23. Karu AE and Belk ED (1982) Induction of *E. coli recA* protein via *recBC* and alternative pathways: quantitation by enzyme-linked immunosorbent assay (ELISA). Mol Gen Genet 185:275-282.
24. Salles B and Paoletti C (1983) Control of UV induction of RecA protein. Proc Natl Acad Sci USA 80:65-69.

25. Moreau PL (1987) Effects of overproduction of single-stranded DNA-binding protein on RecA protein-dependent processes in *Escherichia coli.* J Mol Biol 194:621-634.
26. Meyer RR and Laine PS (1990) The single-stranded DNA-binding protein of *Escherichia coli.* Microbiol Rev 54:342-380.
27. Bobst EV, Bobst AM, Perrino FW, Meyer RR and Rein DC (1985) Variability in the nucleic acid binding site size and the amount of single-stranded DNA-binding protein in *Escherichia coli.* FEBS Letters 181:133-137.
28. Abrams R (1951) Effect of X-rays on nucleic acid and protein synthesis. Arch Biochem 29:90-99.
29. Kelner A (1953) Growth, respiration, and nucleic acid synthesis in ultraviolet-irradiated and in photoreactivated *Escherichia coli.* J Bacteriol 65:252-262.
30. Salles B, Paoletti C and Villani G (1983) Lack of single-strand DNA-binding protein amplification under conditions of SOS induction in *E. coli.* Mol Gen Genet 189:175-177.
31. Perrino FW, Rein DC, Bobst AM and Meyer RR (1987) The relative rate of synthesis and levels of single-stranded DNA binding protein during induction of SOS repair in *Escherichia coli.* Mol Gen Genet 209:612-614.
32. Schuster H, Beyersmann D, Mikolajczyk M and Schlicht M (1973) Prophage induction by high tremperature in thermosensitive *dna* mutants lysogenic for bacteriophage lambda. J Virol 11:879-885.
33. Gudas LJ and Pardee AB (1976) DNA synthesis inhibition and the induction of protein X in *Escherichia coli.* J Mol Biol 101:459-477.
34. Phizicky EM and Roberts JW (1981) Induction of SOS functions: regulation of proteolytic activity of *E. coli* RecA protein by interaction with DNA and nucleoside triphosphate. Cell 25:259-267.
35. Menetski JP and Kowalczykowski SC (1989) Enhancement of *Escherichia coli* RecA protein enzymatic function by dATP. Biochemistry 28: 5871-5881.
36. Munch-Petersen A and Neuhard J (1964) Studies on the acid-soluble nucleotide pool in thymine-requiring mutants of *Escherichia coli* during thymine starvation. I. Accumulation of deoxyadenosine triphosphate in *Escherichia coli* 15 $T^-A^-U^-$. Biochim Biophys Acta 80:542-551.
37. Neuhard J (1966) Studies on the acid-soluble nucleotide pool in thymine-requiring mutants of *Escherichia coli* during thymine starvation. III. On the regulation of the deoxyadenosine triphosphate and deoxycytidine triphosphate pools of *Escherichia coli.* Biochim Biophys Acta 129:104-115.
38. Neuhard J and Thomassen E (1971) Turnover of the deoxyribonucleoside triphosphates in *Escherichia coli* 15T during thymine starvation. Eur J Biochem 20:36-43.
39. Mathews CK (1972) Biochemistry of deoxyribonucleic acid-defective amber mutants of bacteriophage T4. J Biol Chem 247:7430-7438.

40. Filpula D and Fuchs JA (1977) Regulation of ribonucleoside diphosphate reductase synthesis in *Escherichia coli*: increased enzyme synthesis as a result of inhibition of deoxyribonucleic acid synthesis. J Bacteriol 130:107-113.
41. Thelander L and Reichard P (1979) Reduction of ribonucleotides. Ann Rev Biochem 48:133-158.
42. Bochner BR and Ames BN (1982) Complete analysis of cellular nucleotides by two-dimensional thin layer chromatography. J Biol Chem 257:9759-9769.
43. Neuhard J and Nygaard P (1987) Purines and pyrimidines. In: Neidhardt FC, ed. *Escherichia coli* and *Salmonella typhimurium*. Cellular and molecular biology. Washington, D.C.: American Society for Microbiology, pp 445-473.
44. Barbé J, Villaverde A and Guerrero R (1983) Evolution of cellular ATP concentration after UV-mediated induction of SOS system in *Escherichia coli*. Biochem Biophys Res Commun 117:556-561.
45. Guerrero R, Llagostera M, Villaverde A and Barbé J (1984) Changes in ATP concentration in *Escherichia coli* during induction of the SOS system by mitomycin C and bleomycin. J Gen Microbiol 130:2247-2251.
46. Barbé J, Vericat JA, Cairó J and Guerrero R (1985) Further characterization of SOS system induction in *recBC* mutants of *Escherichia coli*. Mutat Res 146:23-32.
47. Kuzminov A (1995) A mechanism for induction of the SOS response in *E. coli*: insights into the regulation of reversible protein polymerizaton in vivo. J Theor Biol 177:29-43.
48. Lusk JE, Williams JP and Kennedy EP (1968) Magnesium and the growth of *Escherichia coli*. J Biol Chem 243:2618-2624.
49. Günther T and Dorn F (1969) Uber die intrazellulare Mg-ionenaktivitat von *E. coli*-zellen. Z Naturforschg 24b: 713-717.
50. Alatossava T, Jütte H, Kuhn A and Kellenberger E (1985) Manipulation of intracellular magnesium content in polymyxin B nonapeptide-sensitized *Escherichia coli* by ionophore A23187. J Bacteriol 162:413-419.
51. Hurwitz C and Rosano CL (1967) The intracellular concentration of bound and unbound magnesium ions in *Escherichia coli*. J Biol Chem 242:3719-3722.
52. Kowalczykowski SC and Krupp RA (1987) Effects of *Escherichia coli* SSB protein on the single-stranded DNA dependent ATPase activity of *Escherichia coli* RecA protein. Evidence that SSB protein facilitates the binding of RecA protein to regions of secondary structure within single-stranded DNA. J Mol Biol 193:97-113.
53. Lavery PE and Kowalczykowski SC (1990) Properties of RecA441 protein-catalyzed DNA strand exchange can be attributed to an enhanced ability to compete with SSB protein. J Biol Chem 265:4004-4010.
54. Thresher RJ, Christiansen G and Griffith JD (1988) Assembly of presynaptic filaments. Factors affecting the assembly of RecA protein onto single-stranded DNA. J Mol Biol 201:101-113.

55. Lowry OH, Carter J, Ward JB and Glaser L (1971) The effect of carbon and nitrogen sources on the level of metabolic intermediates in *Escherichia coli*. J Biol Chem 246:6511-6521.
56. Ferrari ME, Bujalowski W and Lohman TM (1994) Cooperative binding of *Escherichia coli* SSB tetramers to single-stranded DNA in the $(SSB)_{35}$ binding mode. J Mol Biol 236:106-123.
57. Bujalowski W and Lohman TM (1986) *Escherichia coli* single-strand binding protein forms multiple, distinct complexes with single-stranded DNA. Biochemistry 25:7799-7802.
58. Wei T-F, Bujalowski W and Lohman TM (1992) Cooperative binding of polyamines induces the *Escherichia coli* single-strand binding protein-DNA binding mode transition. Biochemistry 31:6166-6174.
59. Lohman TM, Bujalowski W and Overman LB (1988) *E. coli* single-strand binding protein: a new look at helix-destabilizing proteins. Trends Biochem Sci 13:250-255.
60. Curth U, Greipel J, Urbanke C and Maass G (1993) Multiple binding modes of the single-stranded DNA binding protein from *Escherichia coli* as detected by tryptophan fluorescence and site-directed mutagenesis. Biochemistry 32:2585-2591.
61. Carlini LE, Porter RD, Curth U and Urbanke C (1993) Viability and preliminary in vivo characterization of site-directed mutants of *Escherichia coli* single-stranded DNA-binding protein. Mol Microbiol 10:1067-1075.
62. Griffith JD, Harris LD and Register III J (1984) Visualization of SSB-ssDNA complexes active in the assembly of stable RecA-DNA filament. Cold Spring Harbor Symp Quant Biol 49:553-559.
63. Kowalczykowski SC (1991) Biochemistry of genetic recombination: Energetics and mechanism of DNA strand exchange. Annu Rev Biophys Biophys Chem 20:539-575.
64. Muniyappa K, Williams K, Chase JW and Radding CM (1990) Active nucleoprotein filaments of single-stranded binding protein and RecA protein on single-stranded DNA have a regular repeated structure. Nucleic Acids Res 18:3967-3973.
65. Cotterill SM, Satterthwait AC and Fersht AR (1982) RecA protein from *Escherichia coli*. A very rapid and simple purification procedure: binding of adenosine 5'-triphosphate and adenosine 5'-diphosphate by the homogeneous protein. Biochemistry 21:4332-4337.
66. Cotterill SM and Fersht AR (1983) RecA filaments in solution. Biochemistry 22:3525-3531.
67. Griffith J and Shores CG (1985) RecA protein rapidly crystallizes in the presence of spermidine: a valuable step in its purification and physical characterization. Biochemistry 24:158-162.
68. Shibata T, Cunningham RP and Radding CM (1981) Homologous pairing in genetic recombination. Purification and characterization of *Escherichia coli* RecA protein. J Biol Chem 256:7557-7564.
69. Tsang SS, Muniyappa K, Azhderian E, Gonda DK, Radding CM and Flory J (1985) Intermediates in homologous pairing promoted by RecA

protein. Isolation and characterization of active presynaptic complexes. J Mol Biol 185:295-309.
70. Lavery PE and Kowalczykowski SC (1988) Biochemical basis of the temperature-inducible constitutive protease activity of the RecA441 protein of *Escherichia coli*. J Mol Biol 203:861-874.
71. Muniyappa K, Shaner SL, Tsang SS and Radding CM (1984) Mechanism of the concerted action of recA protein and helix-destabilizing proteins in homologous recombination. Proc Natl Acad Sci USA 81: 2757-2761.
72. Chabbert M, Cazenave C and Helene C (1987) Kinetic studies of RecA protein binding to a fluorescent single-stranded polynucleotide. Biochemistry 26:2218-2225.
73. Horii T, Ogawa T, Nakatani T, Hase T, Matsubara H and Ogawa H (1981) Regulation of SOS functions: purification of *E. coli* LexA protein and determination of its specific site cleavage by the RecA protein. Cell 27:515-522.
74. Little JW, Mount DW and Yanisch-Perron CR (1981) Purified LexA protein is a repressor of the *recA* and *lexA* genes. Proc Natl Acad Sci USA 78:4199-4203.
75. West SC and Little JW (1984) *P. mirabilis* RecA protein catalyses cleavage of *E. coli* LexA protein and the λ repressor in vitro. Mol Gen Genet 194:111-113.
76. Roberts JW, Roberts CW and Mount DW (1977) Inactivation and proteolytic cleavage of phage λ repressor in vitro in ATP-dependent reaction. Proc Natl Acad Sci USA 74:2283-2287.
77. Shibata T, DasGupta C, Cunningham RP, Williams JGK, Osber L and Radding CM (1981) Homologous pairing in genetic recombination. The pairing reaction catalyzed by *Escherichia coli* RecA protein. J Biol Chem 256:7565-7572.
78. Tsang SS, Chow SA and Radding CM (1985) Networks of DNA and RecA protein are intermediates in homologous pairing. Biochemistry 24:3226-3232.
79. Roman LJ and Kowalczykowski SC (1986) Relationship of the physical and enzymatic properties of *Escherichia coli* RecA protein to its strand exchange activity. Biochemistry 25:7375-7385.
80. Story RM, Weber IT and Steitz TA (1992) The structure of the *E. coli* RecA protein monomer and polymer. Nature 355:318-325.
81. Freitag N and McEntee K (1988) Affinity chromatography of RecA protein and RecA nucleoprotein complexes on RecA protein-agarose columns. J Biol Chem 263:19525-19534.
82. Wilson DH and Benight AS (1990) Kinetic analysis of the pre-equilibrium steps in the self-assembly of recA protein from *Escherichia coli*. J Biol Chem 265:7351-7359.
83. Higashitani N, Higashitani A, Roth A and Horiuchi K (1992) SOS induction in *Escherichia coli* by infection with mutant filamentous phages that are defective in initiation of complementary-strand DNA synthesis. J Bacteriol 174:1612-1618.

84. Gigliani F, Ciotta C, Del Grosso MF and Battaglia PA (1993) pR plasmid replication provides evidence that single-stranded DNA induces the SOS system in vivo. Mol Gen Genet 238:333-338.
85. De Mot R, Laeremans T, Schoofs G and Vanderleyden J (1993) Characterization of the *recA* gene from *Pseudomonas fluorescens* OE 28.3 and construction of a *recA* mutant. J Gen Microbiol 139:49-57.
86. Williams KR, Spicer EK, LoPresti MB, Guggenheimer RA and Chase JW (1983) Limited proteolysis studies on the *Escherichia coli* single-stranded DNA binding protein. Evidence for a functionally homologous domain in both the *Escherichia coli* and T4 DNA binding proteins. J Biol Chem 258:3346-3355.
87. Egner C, Azhderian E, Tsang SS, Radding CM and Chase JW (1987) Effects of various single-stranded-DNA-binding proteins on reactions promoted by RecA protein. J Bacteriol 169:3422-3428.
88. Cheo DL, Bayles KW and Yasbin RE (1991) Cloning and characterization of DNA damage-inducible promoter regions from *Bacillus subtilis*. J Bacteriol 173:1696-1703.
89. Hertman I and Luria SE (1967) Transduction studies on the role of a *rec*$^+$ gene in the ultraviolet induction of prophage lambda. J Mol Biol 23:117-133.
90. Witkin EM (1991) RecA protein in the SOS response: milestones and mysteries. Biochimie 73:133-141.

CHAPTER 4

Daughter-Strand Gap Repair

4.1. ORIENTATION: THE ORIGIN OF DAUGHTER-STRAND GAPS AND THEIR REPAIR

When *E. coli* cells are irradiated with UV, the molecular weight of their native chromosomal DNA is not lowered. However, when DNA from irradiated cells is denatured with alkali, it has a lower molecular weight than denatured DNA from unirradiated control cells, indicating the presence of single-strand interruptions throughout the irradiated chromosome. Most of these interruptions are due to the ongoing excision repair of pyrimidine dimers, since in excision repair-deficient mutants the molecular weight of the irradiated chromosome is almost the same as the one of intact chromosome. However, the *newly-synthesized DNA* has single-strand interruptions even in excision repair-deficient mutants.

The single-strand interruptions in the newly synthesized DNA after UV-irradiation are due to replisome encounters with the lesions. When a replisome encounters a noncoding lesion (a pyrimidine dimer, or an abasic site) in a template DNA strand, it stops, because DNA polymerases are allowed to synthesize across such lesions only in special circumstances (4.6). Replication of both strands in *E. coli* chromosome is believed to be discontinuous (1.3.2), and the other replisomes are likely to restart downstream as scheduled, but the DNA segment between the site of lesion and the replication restart will be left single-stranded (Fig. 4.1B). Since such a single-strand gap forms in only one of the two daughter branches of the replicating chromosome, it is called a daughter-strand gap (DSG). Daughter-strand gaps were first detected and explained by Rupp and Howard-Flanders,[1] and their length was determined to be on average 800 nucleotides,[2] which is approximately half the average length of the Okazaki pieces in *E. coli*.[3] However, recently with a different method, an estimate of 100-200 nucleotides for the size of DSG in both the leading and the lagging strands in particular DNA sequences was obtained,[4] indicating that the gaps might be quite small.

One way to fill a daughter-strand gap would be to modify a stalled replisome so as to allow it to carry out translesion DNA synthesis

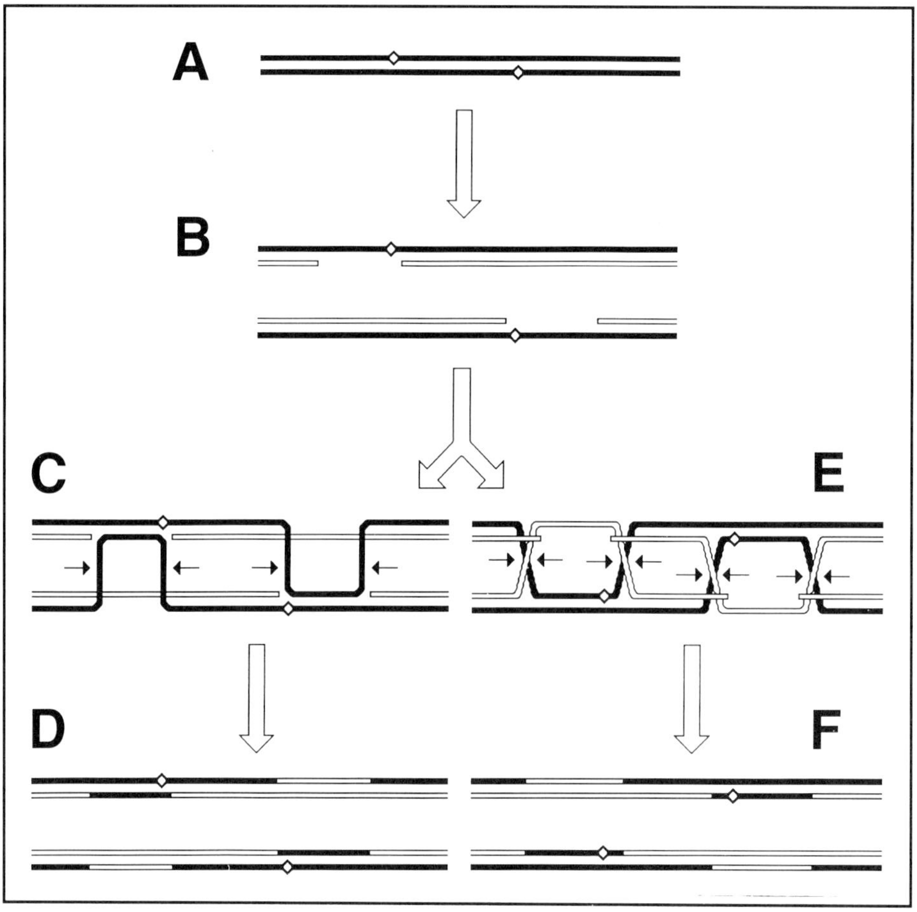

Fig. 4.1. Experimentally-established principles of daughter-strand gap repair in E. coli. *Filled lines – old DNA strands, open lines – newly-synthesized DNA strands, diamonds – noncoding lesions, small arrows – single-strand scissions required to resolve the joint molecules. (A) A DNA molecule containing unrepaired noncoding lesions (for example, thymine dimers). (B) Replication of this molecule will produce two molecules with single-strand gaps in daughter strands opposite the lesions.[1] Eventually, these gaps are repaired in a RecA-dependent way.[7] Experiments of Rupp, Howard-Flanders and colleagues[5] had revealed that after gap closure, the newly-synthesized strands are connected with the old strands (strand exchange), suggesting a model for recombinational repair of daughter-strand gaps (C and D). Later, Ganesan[8] had found that the lesions are frequently transferred from the old strands into the newly-synthesized strands, suggesting that the old strands containing the lesions are exchanged (E and F).*

(4.6). However, experiments of Rupp, Howard-Flanders and colleagues revealed a peculiar feature of daughter-strand gap repair in *E. coli*: filling in the gaps was accompanied by formation of a hybrid DNA between the newly-synthesized and template DNAs.[5] The number of such exchanges of strands between sister duplexes roughly coincided with the number of UV-lesions in template DNA, indicating that daughter-strand gap repair is accompanied by strand exchange (Fig. 4.1C and D). This phenomenon led the authors to propose that daughter-strand gaps in *E. coli* are filled with a help of intact sister duplexes by recombinational repair.[5,6] This conclusion was in line with the finding that daughter-strand gaps are not repaired in *recA* mutants deficient in homologous recombination.[7,8] At last, it was demonstrated that during daughter-strand gap repair, the initial lesions in the old DNA strands have 50% probability to be transferred to the newly-synthesized DNA strands suggesting formation of Holliday junction(s) (1.1.4) and their alternative resolution[8] (Fig. 4.1E and F).

In *E. coli*, daughter-strand gaps are mended by the RecF-pathway of recombinational repair, which is named after the first discovered gene specific for this pathway.[9,10] The complete repair team consists of several general DNA maintenance proteins, including DNA gyrase, DNA topoisomerase I and PriA, which participate not only in DSG repair and other recombinational repair pathways, but also in DNA replication. Besides the main recombinational activator of the RecF pathway, RecA protein, there are other important players including RecF, RecO, RecR, RuvA, RuvB and RuvC proteins. Mutants in some of the corresponding genes, as well as *recA* mutants, are deficient in filling-in the daughter-strand gaps, which accounts for their UV-sensitivity. However, mutants in the other genes fill in gaps normally, but are UV-sensitive anyway, apparently due to the block at a later stage of the repair reaction.

4.2. PRESYNAPTIC PHASE OF DSG-REPAIR: RECF, RECO AND RECR PROTEINS

To be repaired, a single-strand gap-containing duplex needs to be synapsed with an intact homologous duplex. The synapsis reaction is catalyzed by RecA filament (2.2), which must assemble on the single-strand gap. However, single-strand DNA at the gap is complexed with SSB (it was complexed with SSB when it was made single-stranded at the replication fork), and RecA polymerization on ssDNA is inhibited by SSB under normal conditions (3.2). Thus, an activity is required to help RecA to polymerize at the single-stranded gap in the presence of SSB. A combination of genetic and biochemical evidence strongly suggests that RecF, RecO and RecR proteins sense the stalled replication fork and begin daughter-strand gap repair by helping RecA protein to form a filament on a ssDNA covered with SSB.

4.2.1. *RECF*, *RECO* AND *RECR*: GENE STRUCTURE

Genetic analysis of UV-resistance in *E. coli* shows that *recF*, *recO* and *recR* genes belong to the same epistasis group, that is, the double mutants in these genes have the same survival after UV-irradiation as single mutants do.[11,12] Moreover, the mutants in all three genes are deficient in filling-in the daughter-strand gaps (ref. 13 and references therein), although they have no effect on repair of double-strand breaks (5.3 and 5.4). The UV-sensitivity of all these mutants is partially suppressed by the same mutations in *recA* gene.[14-16] Since filling-in the gaps is likely to follow closely RecA-mediated pairing, the failure of *recF*, *recO* and *recR* mutants to fill-in the gaps argues for their action before or during the RecA-promoted step with a possibility of direct interactions between these proteins and RecA.

recF is the third gene in a four-gene cluster, which also includes *dnaA* (DnaA protein initiates chromosomal DNA replication at *oriC*), *dnaN* (a gene coding for the "sliding clamp" subunit of DNA polymerase III) and *gyrB* (B-subunit of DNA gyrase, the enzyme that introduces negative supercoiling into the *E. coli* chromosome (see below)). Although *recF* has its own promoter inside the coding sequence of *dnaN* gene, it is thought that *dnaA*, *dnaN* and *recF* constitute an operon under the control of *dnaA* promoters.[17] The preserved structure of this region in other, even distantly related bacteria, further argues for the biological significance of this combination of *recF* with replication genes (refs. 18, 19 and references therein). The complex regulation of the *recF* gene expression[17,20,21] ensures that RecF protein is maintained at a low level, less than 190 monomers per cell,[22] which is close to the amount of DnaN dimers (about 150) in the cell.[23] Overproduction of RecF protein adversely affects cell viability and UV-resistance.[24,25]

recO is the last gene in a three-gene operon.[26,27] The first gene in the operon is *rnc*, encoding RNase III which cleaves specific double-stranded RNA structures (the enzyme participates in maturation of rRNA and modifies some mRNAs). The second gene is *era*, coding for an essential cytoplasmic membrane-associated GTPase suspected to participate in membrane signaling.[28] *recO* is poorly expressed due to a weak promoter, a weak ribosome binding site and an abundance of rare codons.[26]

recR is also the last, poorly expressed gene in a three-gene operon. The first gene in this operon is *dnaX*, encoding the gamma and tau subunits of DNA pol III; a function of the second gene in the operon, *orf12*, is still unknown.[29,30] *recR* is expressed equally from both the *dnaX* promoter, and from the second promoter, which it shares with *orf12*, and which is buried in the *dnaX* coding sequence.[30] In *Bacillus subtilis*, the genetic organization of this chromosomal region is similar,[31] arguing against fortuitous association of *recR* and *dnaX*. In prokaryotes, where transcription and translation are coupled in space, the proximity of the genes often signals the interaction between the gene products. The colocalization of *recF* and *recR* genes in operons with the genes coding for components of the replicative DNA polymerase may indicate the physical association of RecF or RecR proteins with DNA polymerase III and/or an interaction between these proteins and DNA polymerase III during the repair process. Moreover, the position of both *recF* and *recR* promoters inside the coding sequences of the genes for DNA pol III subunits suggests that their expression is coregulated.

4.2.2. Properties and Activities of RecF, RecO and RecR Proteins

RecF is a 40 kDa protein which, in the absence of divalent cations and nucleotide cofactors, binds linear ssDNA with some preference for the ends.[24] In the presence of ATP and magnesium RecF binds both ssDNA and dsDNA.[32] In fact, binding to DNA stimulates ATP binding by the protein, although no ATP hydrolysis is detected.[32] ATP binding is important for RecF function, since a mutation in nucleotide-binding domain of the protein results in its inactivation.[33]

RecO is a 26 kDa protein that binds both double-stranded and single-stranded DNA in the presence of magnesium and does not require nucleotide cofactors.[34] It can also promote renaturation of complementary single strands, again independently of nucleotide cofactor. RecR protein of *E. coli* has a calculated MW of 22 kDa, but is not yet characterized biochemically. RecR protein of *Bacillus subtilis* binds both single-stranded and double-stranded DNA in a cooperative and distributive manner, organizing DNA into a sort of a scaffold.[35]

4.2.3. Interaction of RecF, RecO and RecR Proteins with RecA and SSB Proteins In Vitro

When added in vitro to the SSB-containing RecA-promoted strand exchange reaction between circular single-stranded and linear duplex DNA, at low concentrations RecF protein does not show any effect, and at high concentrations it inhibits the reaction.[22,36] This inhibition is not observed if RecO and RecR proteins are also present in the reaction. In contrast to the adverse effect of the RecF presence, RecO and RecR proteins stimulate the RecA-promoted reaction, maximal stimulation being observed at a RecO:RecR ratio of 1:1.[36] Interestingly, RecF is completely dispensable for this stimulation. RecO+RecR proteins stimulate the rate of initiation of joint molecule formation, but do not influence the rate of strand exchange itself.[36] These observations suggest that RecO and RecR proteins, possibly working in a complex, help RecA to displace SSB protein from ssDNA. However, the role of RecF protein remains unclear.

Further studies have shown that RecO can physically interact with both RecR and SSB.[37] Moreover, RecO and RecR bind to the

SSB-covered single-stranded DNA without displacing SSB. This RecO and RecR binding allows more efficient RecA binding to the SSB-covered DNA. Although RecA cannot bind to saturation to ssDNA covered with SSB-RecO-RecR complexes, the resulting RecA filaments are fully proficient in pairing and strand-exchange.[37]

Bacteriophage T4 offers a similar perspective on this presynaptic step of recombinational repair. gp32 (SSB protein of T4) inhibits ssDNA binding of UvsX protein (T4's RecA homolog).[38,39] Another protein, UvsY, binds to gp32-complexed ssDNA to promote the subsequent UvsX polymerization.[39,40] Thus, the role played in *E. coli* by two proteins, RecO and RecR, seems to be performed in T4 by a single UvsY protein. Whether T4 has a RecF analog or not is still unknown.

4.2.4. A Model for RecFOR-Catalysis of RecA Polymerization at Daughter-Strand Gaps

A function of RecF might be to sense a stalled replisome and to bring RecO and RecR proteins to the lingering SSB-covered ssDNA gap. RecF could function as such a sensor if it is bound to every started Okazaki piece on its 5'-side (Fig. 4.2). For example, RecF might be initially associated with the DnaN clamps when they are put on the RNA primers, but would be eventually displaced from DnaN by the replisome and left behind on the 5'-end of the primer. Sitting on the 5' end of the primer, RecF might target RecO and RecR proteins to the junction of single-stranded and duplex DNA. RecOR complex would bind SSB-covered DNA in the vicinity of the transition between single-stranded and duplex DNA and promote RecA filament formation. If the template for the previous Okazaki fragment is intact, the charging replisome would bump RecA, RecF, RecO and RecR proteins, together with SSB protein, off the DNA and complete the Okazaki fragment. However, if the replisome is stalled at a lesion, RecA filament will be stabilized (3.2), will reach the stalled replisome and promote its dissociation (4.3). Similar ideas have been discussed by others.[32,41]

The proposed role for RecF protein is consistent with the observation that a RecF-deficiency is partially suppressed by overproduction of RecR alone.[41] Interestingly, although RecO is required for this suppression, it need not be overproduced. This indicates that of the two proteins, RecR might act as a delivery vehicle for RecO, which then assists RecA in polymerization on the SSB-bound ssDNA. There is an additional indirect evidence supporting this conclusion. A protein assisting in RecA polymerization on SSB-covered DNA is likely to have a prominent positively-charged domain (3.3.3). By deduced amino acid sequence[42] RecR is a weakly acidic protein with charges distributed more-or-less evenly over the entire molecule. In contrast, the deduced amino acid sequence of RecO protein[26] reveals a highly basic protein, with almost the entire surplus positive charge (+10) concentrated in the carboxy-terminal 17% of the protein. Thus, of the two proteins,

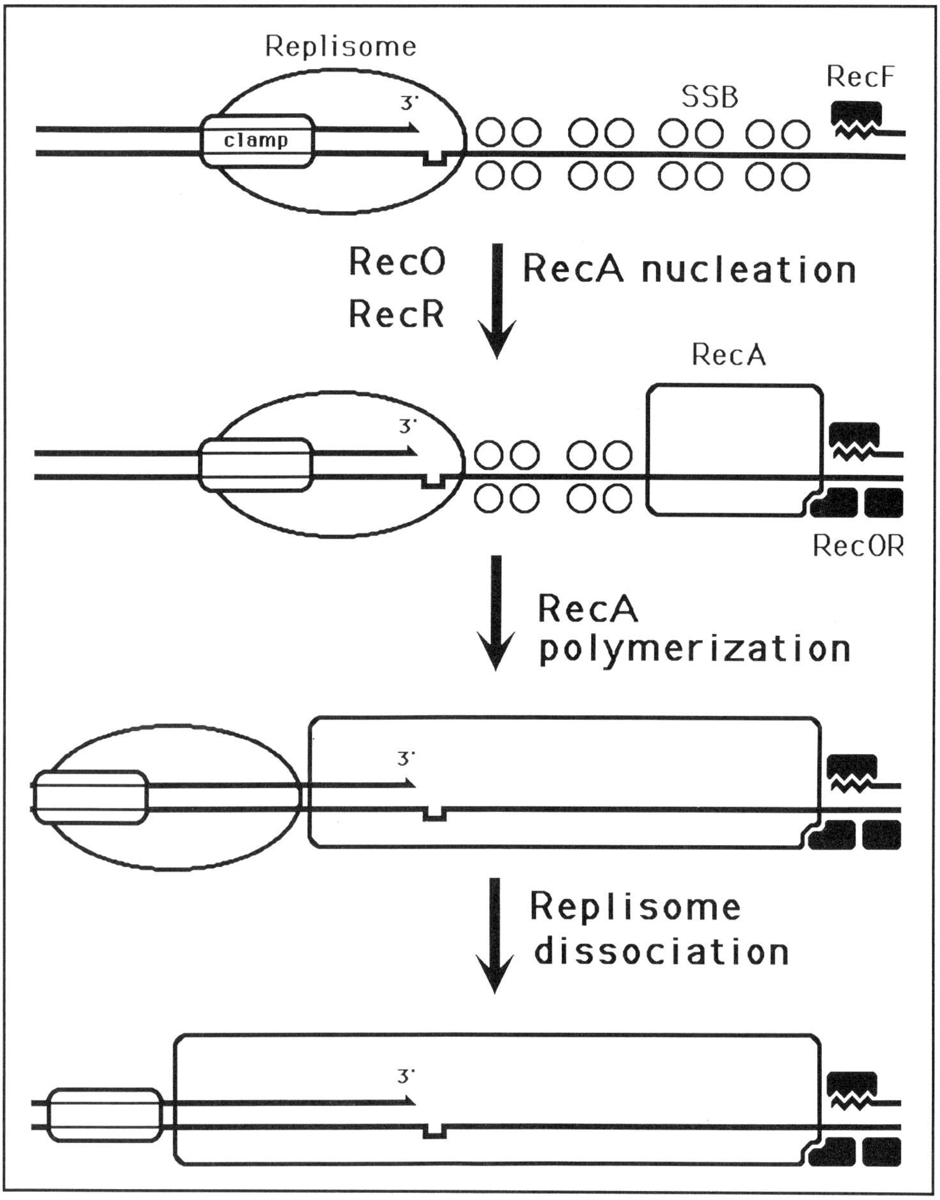

Fig. 4.2. Presynaptic phase of daughter-strand gap repair: RecA filament assembly and replisome reactivation. Replisome is depicted as a big open oval, its DnaN subunit (the clamp) as an almost included open rectangle with rounded corners. The 3' end of the nascent DNA is stalled at a lesion (the irregularity in the lower strand). SSB is shown as open circle quartets around single-stranded region; RecF (black loaf) sits on the RNA primer (wavy segment in the upper strand at the very right). RecA filament is shown as a big open rectangle with rounded corners spreading to the left. Explanations in the picture and in the text.

RecO is more likely to be able to assist with RecA polymerization. In fact, having a highly positively charged tail, RecO should be able to bind RecA which has two negatively-charged domains.[43] Affinity chromatography with RecA protein as a ligand reveals that of all the *E. coli* soluble proteins, only two are retained on the RecA column: RecA itself and another protein with MW about 25 kDa,[44] which is close to the calculated MW for RecO protein (26 kDa). Thus, indirect evidence singles out RecO as the protein which may directly interact with RecA.

4.2.5. The Role of RecFOR in SOS Induction

When replication forks are inhibited by DNA damage, the SOS response is induced (3.1). Under these conditions, the concentration of ATP, dATP, Mg^{2+} and spermidine are believed to be elevated, allowing RecA to form a stable filament on ssDNA in the presence of SSB (3.2). In vitro under similar conditions RecA alone can displace SSB from ssDNA. Why then, in *recF*, *recO* or *recR* mutants, is the SOS response not induced by massive DNA damage, even when the concentration of precursors and counterions should be high enough to permit unassisted SSB displacement by RecA? One explanation may be that the amount of RecA protein present in uninduced cells is inadequate for the repair of numerous daughter-strand gaps. In the absence of RecF (RecO, RecR) protein but in conditions favoring SSB displacement, RecA might form multiple short filaments dispersed among several single-strand gaps. This state does not allow sufficient LexA cleavage to induce the SOS response. RecF, RecO and RecR proteins could allow the assembly of RecA filaments, proficient in LexA cleavage, at the very first daughter-strand gap, when the concentration of precursors and counterions is only starting to increase. By the time the elevated concentration of precursors and counterions favor unassisted RecA polymerization on SSB-bound ssDNA, the amount of RecA would be amplified to a level adequate for the repair of multiple daughter-strand gaps.

According to this explanation, amplification of RecA prior to DNA damage should suppress a *recF* deficiency. RecA amplification could be achieved in several ways: by placing *recA* gene onto a multicopy plasmid, by an up-mutation in the RecA promoter, or by inactivation of LexA repressor. RecA amplification together with other SOS proteins due to defective LexA repressor does slightly suppress a *recF* deficiency.[45] However, overproduction of RecA alone due to a promoter-up mutation actually decreases UV-survival of *recF* mutants[46] and does not improve the slow SOS induction in them.[47] Therefore, RecA amplification is unlikely to be the function of RecFOR in vivo.

An alternative explanation for the role of RecFOR in the SOS response postulates that, in contrast to RecA filament self-assembly in vitro, RecA polymerization in vivo always requires the assistance of other proteins (3.3.3). The function of RecFOR could be to catalyze RecA filament assembly on daughter-strand gaps. The regulation of the SOS response, according to this logic, is achieved on the level of RecA filament stability. Under normal conditions, RecA filaments would be destabilized by SSB and low concentrations of precursors and counterions. However, when DNA synthesis is inhibited, resulting in higher concentrations of precursors and counterions, the stability of RecA filaments could be improved, leading to a significantly amount of LexA cleavage and subsequent induction of the SOS response. Since, according to this idea, no RecA filament assembly on daughter-strand gaps in vivo could happen without the assistance of RecFOR, the SOS response would not be induced in the corresponding mutants even if RecA were overproduced.

4.3. RECA POLYMERIZATION: REPLISOME REACTIVATION

DNA synthesis is inhibited within 5-10 minutes in *E. coli* cells irradiated with sub-lethal doses of UV.[48-51] It resumes shortly afterwards at the stalled replication forks in the WT cells,* but never again in *recA* or *lexA* mutants.[49-51] It is thought that when a replisome stops at a noncoding lesion in a template DNA, it needs to be disengaged from the damaged DNA ("reactivated") to restart downstream.[49] Indeed, it can be calculated that the UV fluency required to stop the bulk of DNA replication in an excision repair-deficient strain[50] induces 6-12 pyrimidine dimers in the stretch of DNA that could be traversed by two replisomes within 5-10 minutes (at the UV dose of 3 J/M^2, assuming 30 pyrimidine dimers per 1 J/M^2 per *E. coli* genome[1] and a DNA replication rate of 500 bp/sec per replisome in a minimal medium[53]). The number of pyrimidine dimers thus encountered is very close to the number of replisomes in an *E. coli* cell (about 10),[54] substantiating the assumption that an encounter with a pyrimidine dimer in vivo not only blocks replisome progress, but also traps the replisome in an unproductive complex with damaged DNA.

In the mutant constitutively producing elevated levels of RecA, the induction of other genes of the SOS response is not required for the replisome reactivation; nevertheless, protein synthesis is needed for optimal reactivation.[49,50] A trivial yet unaddressed explanation for the protein synthesis requirement is that substantially more RecA is needed

** At higher doses of UV, the existing replication forks seem to be abandoned, and replication reinitiates at the origin (for review see refs. 51, 52). This reinitiation can be both DnaA-dependent and DnaA-independent. In the latter case, it may be caused by inhibition of RNaseH, the enzyme degrading RNA strand in RNA:DNA hybrids (for discussion, see ref. 52) and is proposed to be the basis for inducible stable DNA replication.[51]*

for efficient replisome reactivation than is present in the cells producing elevated levels of RecA. Alternatively, full replisome reactivation, besides elevated levels of RecA protein, may require some other, LexA-independent but damage-inducible protein(s). There are several genes known to be induced in response to DNA damage that lack LexA-binding sites in their promoters. These include *uvrC* (coding for a subunit of UvrABC excinuclease),[55] *phr* (coding for DNA photolysase),[56] *dnaA* (a gene for the DNA replication initiation protein),[57] *dnaQ* (a gene for the proofreading subunit of DNA polymerase III),[58] and *dnaN* (encoding the processivity subunit of DNA polymerase III).[58,59] Which one of the five might be relevant for replisome reactivation?

A model of replisome reactivation is yet to be proposed, although a plausible scenario is not difficult to imagine. A clue to the understanding of the mechanism of replisome reactivation is the realization that RecA filament polymerizes on a daughter-strand gap always toward the stalled replisome (Fig. 4.2). Since a RecA filament assembled on ssDNA can readily extend into an adjacent dsDNA region (2.2.1), by doing so the filament could push the stalled replisome back into a double-stranded region. Having been displaced to a completely double-stranded region, the replisome should be able to disassemble on its own, since this ability is critical for the replisome cycling from completed Okazaki pieces to new primers. In vitro experiments indicate that replisomes will not cycle from duplex DNA unless provided with an excess of DnaN protein.[60-62] This is consistent with the finding that in vivo, there are 300 DnaN molecules (150 DNA clamps?) per cell,[23] which is at least a 10-fold molar excess over the DNA polymerase III holoenzyme.[54] Since DnaN is associated with primers, and there are likely to be more unused primers when replication is inhibited, an increased requirement for DnaN protein is anticipated by the model. Therefore, DnaN might be one of the additional proteins needed for the efficient replisome reactivation.

According to the suggested model, RecA reactivates replisomes simply by polymerizing on ssDNA before promoting any further reaction (Fig. 4.2). If this idea is correct, then replisome reactivation should be observed in *recA* mutants defective for strand exchange, but proficient in filament assembly. No such mutants have been isolated yet.

4.4. TOPOISOMERASE REQUIREMENTS OF THE SYNAPSIS REACTION: DNA GYRASE AND TOPOISOMERASE I

RecA filament assembled on ssDNA finds a homologous region on the sister chromosome and pairs the ssDNA with this region. Since double-stranded DNA is a helix in which single strands are intertwined around each other, the pairing of the incoming strand with an intact duplex DNA entails two different topological reactions. First, the positive supercoiling generated when the two strands of the intact duplex are pulled apart to accommodate the invading strand must be neutralized (Fig. 4.3). Since the *E. coli* chromosome is normally negatively supercoiled, this results in an initial loss and subsequent restoration of the usual state.

The maintenance of negative supercoiling in the *E. coli* chromosome is a function of DNA gyrase. DNA gyrase mutants are significantly impaired in their ability to perform recombinational repair of UV-induced DNA damage.[63,64] DNA gyrase is a heterotetramer (A_2B_2) of two subunits: GyrA (MW 97 kDa) and GyrB (MW 90 kDa). The subunits are encoded by two separate genes, *gyrA* and *gyrB*, which in many bacteria are organized into an operon, but in *E. coli* are situated in different parts of the chromosome (reviewed in ref. 65). GyrA is a catalytic subunit which binds and manipulates DNA, while GyrB is an ATPase, which binds and manipulates a GyrA-DNA complex. The complicated mechanism of DNA gyrase action has been elucidated in detail (reviewed in ref. 65). In a nutshell, the enzyme binds two adjacent regions of the same duplex DNA molecule, makes a staggered double-strand cut in one of them, passes the uncut region through the resulting gap and reseals the gap. As a result, the number of times one strand is wound around the other in the DNA molecule is decreased by two, introducing two negative supercoils in the molecule.

The incoming single strand of the damaged duplex forms a homologous complex with the complementary strand of the intact duplex. At first, this homologous complex is paranemic in nature, meaning that the two strands are not truly intertwined around each other (Fig. 4.3). In other words, they have so many negative supercoils that they can be freely separated from each other. To give the natural conformation to the nascent double-stranded DNA region, a topological reaction opposite to the one catalyzed by DNA gyrase must occur. The paranemic duplex is converted to a plectonemic (intertwined) duplex when the excess negative supercoils are removed in a reaction catalyzed in *E. coli* by DNA topoisomerase I (Fig. 4.3).

Mutants in DNA topoisomerase I are sensitive to UV light,[66,67] although the most affected stage of the repair still remains to be determined. DNA topoisomerase I is a single polypeptide with MW 110 kDa, encoded by *topA* gene.[68] Since the relaxation of negative supercoiling in DNA is a restoration of the energetically-favored conformation, the enzyme does not need high-energy cofactors. The mechanism of its reaction is fundamentally different from that of DNA gyrase. DNA topoisomerase I of *E. coli* cuts only one DNA strand, preserves the energy of the phosphodiester bond by covalently attaching itself to the 5' phosphate of the break, rotates one side of the break around the intact strand, and reseals the break. One such manipulation reduces the number of negative supercoils in a DNA molecule by one.

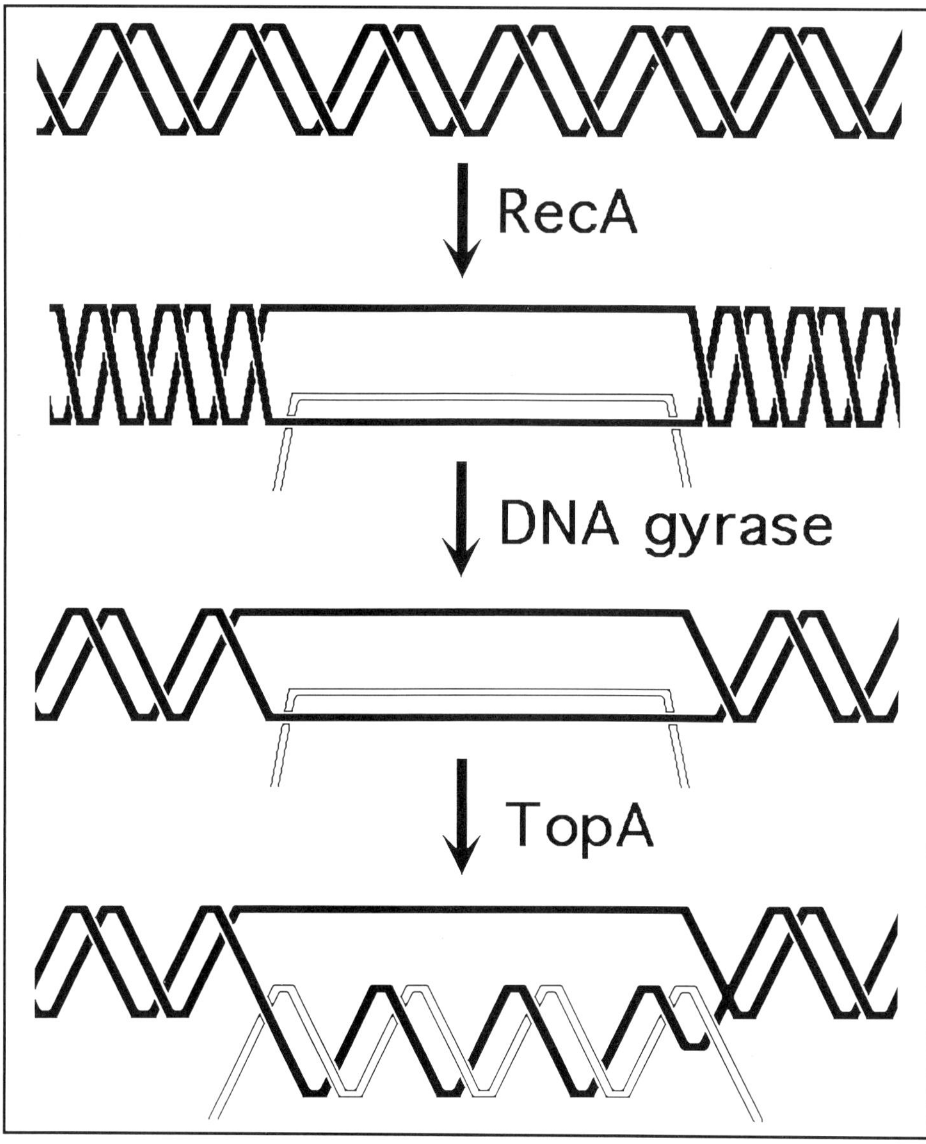

Fig. 4.3. Synaptic phase: topoisomerase requirements. Regular DNA superhelicity is shown in the top and in the bottom helices. In the two middle panels the helices are either overwound (too much supercoiling) or underwound (complete absence of supercoiling). DNA gyrase removes extra coils from the black-black duplex, while DNA topoisomerase I introduces missing coils in the black-white duplex.

4.5. POSTSYNAPTIC PHASE OF DSG-REPAIR: RuvA, RuvB AND RuvC PROTEINS

As RecA catalyzes invasion of the single-stranded region of a gapped DNA molecule into an intact homologous duplex, a pair of branch points at which the invading strand trades places with the homologous resident strand is formed. RecA filament assembles on ssDNA in one direction (toward the lesion that caused the replication block) and probably spreads into the neighboring duplex DNA region (2.2.1). Thus, it is likely to drive the branch point with the growing end of the filament, converting the corresponding three-stranded junction into a four-stranded junction (Fig. 4.4). The four-stranded junctions are better known as Holliday junctions, after Robin Holliday, who first envisioned them in his famous scheme for meiotic recombination[69] (1.1.4).

When synapsis of the damaged and the intact DNA sequences is completed, and the RecA filament treadmills away, the DNA irregularities can be fixed by one-strand repair systems. The lesion that once blocked DNA replication is now in the duplex region and can be removed by the excision repair using the complementary strand as a template (1.1.1). The 3' end of the single strand gap, being paired with an intact strand, can prime filling-in of the gap (Fig. 4.4). Daughter-strand gaps are likely to be filled in by DNA polymerase I, since in *polA* mutants, there are no *recA*-dependent long patches after UV-irradiation.[70] When the gap is filled, converting the other three-strand junction into a Holliday junction, the two intact chromosomes are held together by a pair of Holliday junctions with RecA filament in between. The removal of used RecA filaments and resolution of Holliday junctions in *E. coli* is thought to be a function of RuvA, RuvB and RuvC proteins.

4.5.1. *RUV* LOCUS: GENETICS AND MOLECULAR BIOLOGY

Certain mutations conferring sensitivity to mitomycin C and UV light were mapped to a locus called *ruv*.[71,72] *ruv* mutants repair daughter-strand gaps normally and are able to reinitiate DNA replication after the repair, but fail to resume cellular division, forming long nonseptate, multinucleate filaments.[71,72] The filamentation phenotype of *ruv* mutants can be mutationally suppressed without improving the cell's resistance to UV irradiation,[72,73] arguing against the idea[71] that *ruv* mutants are deficient in some function needed for reinitiation of cell division after DNA damage. *ruv* mutants are deficient for conjugative recombination in *recBC sbc* genetic backgrounds (5.6), and for plasmid recombination in a wild type background.[72,74] In plasmid-by-chromosome recombination, the product of the *ruv* locus acts after the *recA* gene product.[75] These data suggest that the product of the *ruv* locus participates in recombinational repair and acts after the formation of joint molecules.

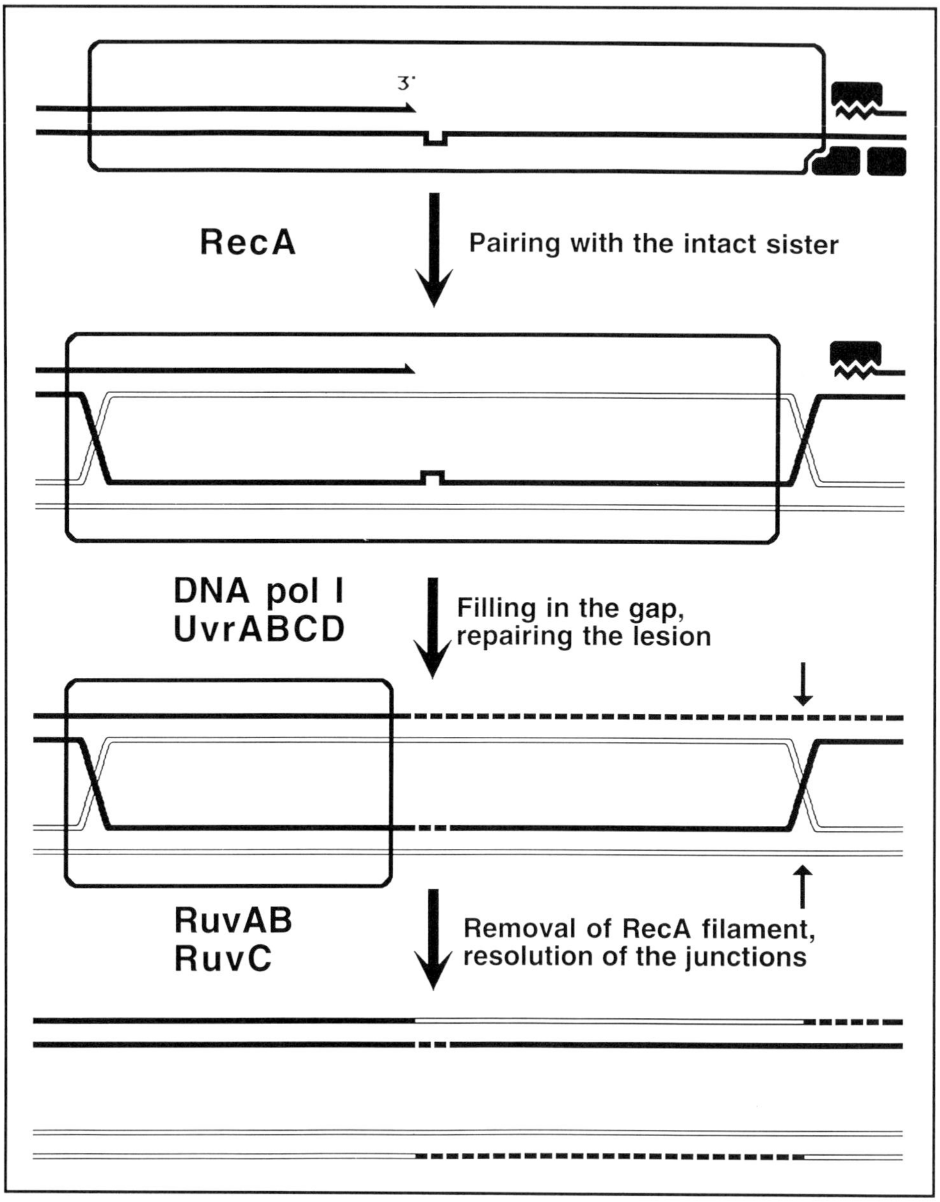

Fig. 4.4. Synaptic and postsynaptic phases of daughter-strand gap repair: pairing with an intact homologue, repair of the lesion and the gap, removal of the used RecA filament and the Holliday junctions. RecA filament is shown as a rectangle with rounded corners which shrinks leftward. Only resolution of the right Holliday junction is indicated by small vertical arrows. Explanations in the figure and in the text.

The *ruv* locus was shown to be induced during the SOS response.[76] Molecular characterization of the locus revealed the presence of three genes. *ruvA* and *ruvB* are organized into an SOS-inducible operon, while *ruvC* is a member of an adjacent, noninducible operon.[77-79] Mutations in any one of the three *ruv* genes confer the same phenotype.[74,79]

Characterization of the biochemical activities of the *ruv* gene products became possible after methods of constructing Holliday junctions in vitro were developed. A model Holliday junction can be assembled with four oligonucleotides.[80] To ensure stability of the junction, each half of a participating oligonucleotide is complementary only to the corresponding half of one of the other oligonucleotides. Such a Holliday junction, called a nonhomologous junction, is unable to do even a limited strand exchange (is "frozen") (Fig. 4.5A). Nonhomologous junctions are substrates for certain cleaving enzymes, but are stable in crude *E. coli* extracts.[81] To imitate the in vivo situation, a short homologous core is introduced in the center of a junction (Fig. 4.5B), enabling it to exchange strands or "branch-migrate" in its central part (Fig. 4.5C and D). Such junctions are either "resolved" (Fig. 4.5E) or "torn apart" (Fig. 4.5F) by recombinational repair enzymes of *E. coli.* The second popular in vitro substrate containing Holliday junctions is an intermediate of RecA-promoted four-strand exchange reaction (Fig. 4.5G)

4.5.2. RuvA and RuvB Proteins: Physical Properties and Interaction with Linear DNA

RuvA protein (MW 22 kDa) behaves like a tetramer or a pentamer in solution which appears to be stabilized by disulfide bridges.[82,83] At high protein concentrations, it can bind single stranded as well as duplex DNA independently of nucleotide cofactors or Mg^{2+},[84] the binding being inhibited at Mg^{2+} concentrations in excess of 5 mM.[85]

RuvB protein (MW 37 kDa) behaves like a dimer in solution[82,83] and is a weak ATPase, unable to bind to ss- or dsDNA at low Mg^{2+} concentrations.[85,86] RuvB starts binding duplex DNA when the Mg^{2+} concentration is raised to 15 mM; its DNA-dependent ATPase activity is stimulated accordingly.[85]

The DNA-stimulated ATPase activity of RuvAB complex (4.5.3) is a hallmark of DNA helicases, the enzymes that catalyze separation of strands (unwinding) in duplex DNA.[87] Indeed, the DNA unwinding activity of RuvAB complex has been demonstrated.[88] EM observations of RuvB in solution with circular DNA show that it forms hexameric rings which interact with duplex DNA as beads on a string.[89] This arrangement of RuvB subunits is reminiscent of several other helicases, for which the same hexameric structure was previously described (references cited in ref. 89).

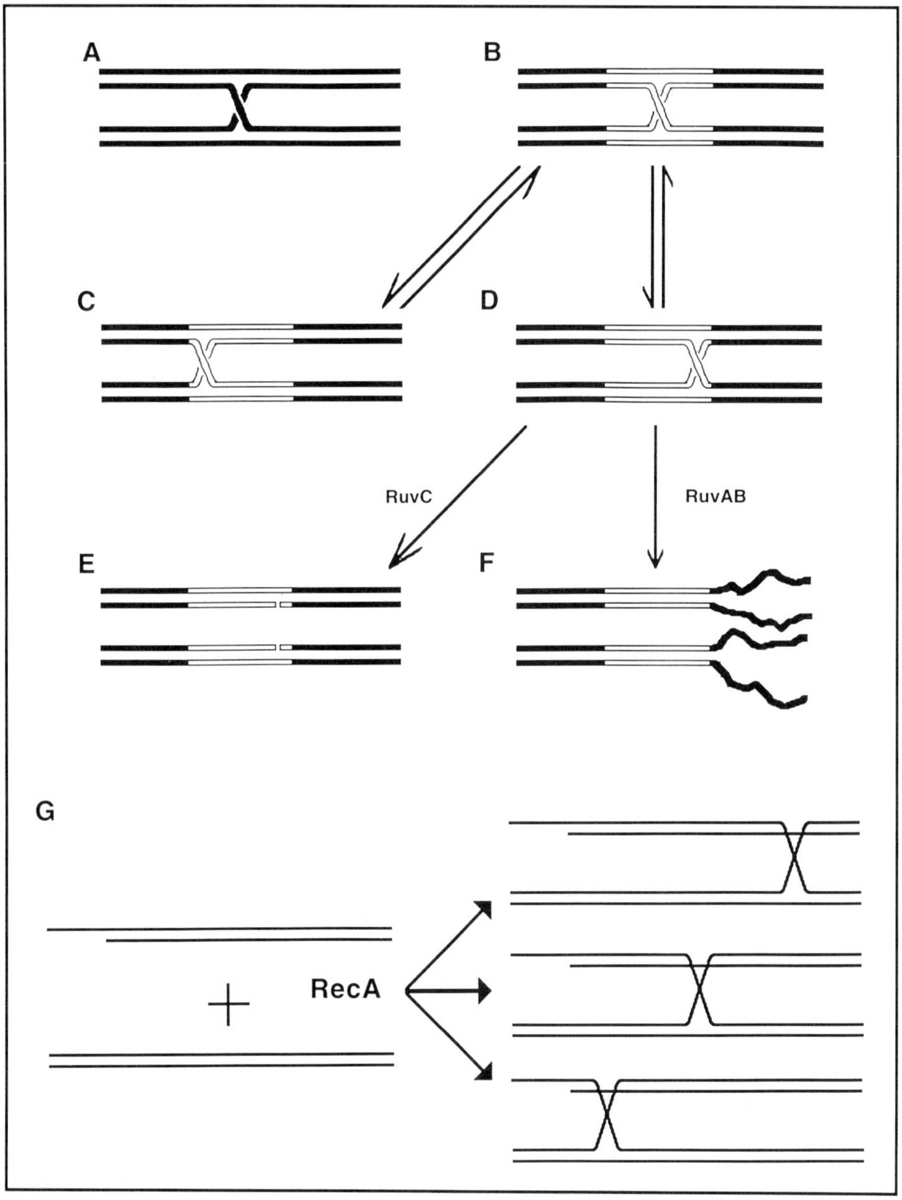

Fig. 4.5. Two ways of constructing model Holliday junctions in vitro. (A–F) Junctions assembled with four oligonucleotides; G, joint molecules, whose formation is catalyzed by RecA protein. (A) A frozen junction, in which the junction point can not migrate due to the absence of homology. (B) A junction with a central homologous core, which allows the junction point to migrate in both directions (C and D). (E) Resolution of such a junction by RuvC. (F) Removal of the junction by RuvAB. (G) A RecA-catalyzed reaction of two homologous linear DNAs, one of them having a 5'-single-stranded overhang, produces a set of joint molecules, each one with a Holliday junction.

4.5.3. RuvAB Interaction with Holliday Junctions

RuvA binds more strongly to Holliday junctions than to linear DNA.[90-92] RuvA and RuvB proteins together bind to a Holliday junction in the presence of a nonhydrolysable ATP analog when Mg^{2+} concentrations are 5 mM or higher.[92] In the presence of RuvA and a Holliday junction, the ATPase activity of RuvB is stimulated at least 10-fold.[84,92] RuvAB complex can hydrolyze several nucleotide cofactors. The maximal rate of hydrolysis is attained with dATP rather than with rATP,[92] which parallels the requirements of RecA protein (ref. 93 and references therein). In the presence of a hydrolysable nucleotide cofactor, RuvAB complex promotes dissociation of artificially produced recombinational intermediates held together by Holliday junctions by moving the junctions to the ends of the linear substrate. RuvAB-promoted Holliday junction migration is observed only when Mg^{2+} concentrations are 5 mM or higher, with an optimal concentration of 10-15 mM,[94] which also parallels the Mg^{2+} requirements for the RecA-promoted strand exchange.[95,96]

In vitro, at physiological Mg^{2+} concentrations, Holliday junctions assume a stacked conformation[97,98] (Fig. 4.6A), impeding spontaneous branch migration.[99] RuvAB might promote migration of "stacked" Holliday junctions using its DNA unwinding activity. For this to happen, two RuvB hexamers bound to two opposite arms of the junction should move away from the junction, unwinding DNA as they move, the unwound complementary strands from different homologous sequences reannealing behind (may be with the help of RuvA?). The net result of this process would be migration of the Holliday junction *behind* RuvB rings (Fig. 4.6B,C).

However, in vitro studies indicate that the mechanism of branch migration by RuvAB may be very different. In solution, when Mg^{2+} concentrations are lowered below a certain level, junctions isomerize to assume an open planar conformation[97] (Fig. 4.6D) allowing rapid spontaneous branch migration.[99] It was suggested, therefore, that Holliday junction-driving enzymes could facilitate branch migration in vivo by decreasing effective Mg^{2+} concentration around the junction.[99] Indeed, it was recently found that binding of RuvA to a junction even in the presence of Mg^{2+} forces it to assume the square planar conformation, as if in the absence of Mg^{2+} ions.[100] This isomerization of the junctions by RuvA could facilitate their subsequent branch-migration by RuvB. Unexpectedly, the direction of such migration is opposite to the direction predicted on the basis of RuvB-as-a-helicase scheme: instead of dragging them behind, RuvB double hexamer seems to push the junctions *ahead* of itself[101] (Fig. 4.6E,F).

Paradoxically, by pushing a Holliday junction ahead of itself, RuvB apparently acts as an annealing enzyme rather than a duplex-unwinding enzyme. Indeed, it should combine strands from duplex arms, separated at the junction, into a new pair of duplex strands. Since RuvAB complex does show a helicase activity, it might be proper to call this annealing activity a "reverse helicase", because its mode of action is likely to be different from the mechanisms of annealing promoted by other proteins. The limited unwinding ability of RuvAB helicase (only

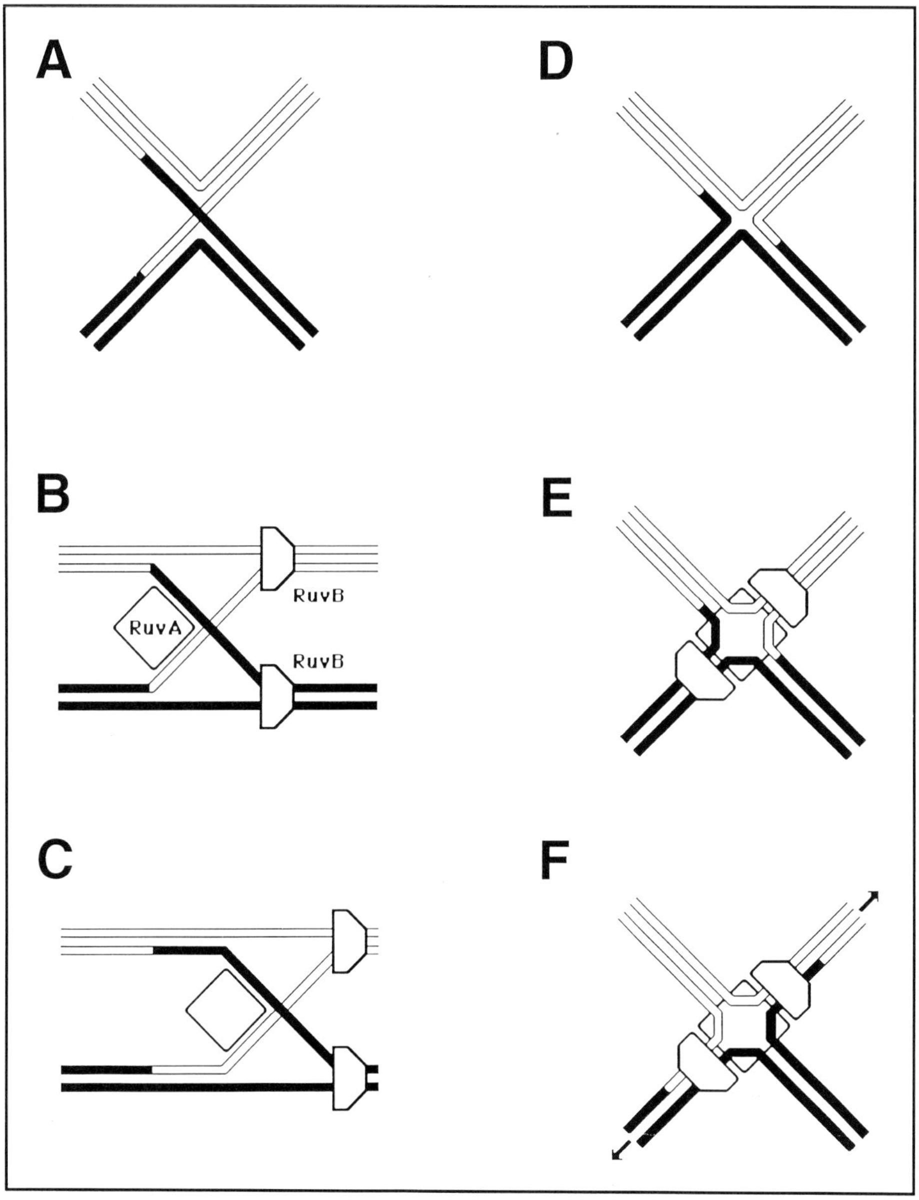

Fig. 4.6. Interaction of RuvA and RuvB proteins with Holliday junctions. RuvAB tetramer is shown as a diamond, RuvB hexameric rings – as trapezoid washers on DNA duplexes. The direction of DNA movement through RuvAB complex in F is shown by arrows. Explanations in the text.

10% of 534 bp duplex is unwound in vitro[88]) together with a possible transient nature of unwound DNA within RuvAB complex[102] are consistent with the idea that RuvAB enzyme acts as a "reverse helicase" when moving Holliday junctions.

4.5.4. RuvAB Enzyme Removes RecA Filament from Duplex DNA

Why is migration of Holliday junctions important for completion of recombinational repair? The phenotype of *ruvC* mutants is indistinguishable from the phenotypes of *ruvA* or *ruvB* mutants,[72,74] suggesting that translocation of Holliday junctions by RuvAB is required for their resolution by RuvC (additional evidence substantiating this idea will be presented later). The other potential role for junction migration could be depolymerization of RecA filaments.

Heavy damage to chromosomal DNA causes formation of single-stranded regions in DNA and inhibits DNA synthesis. Inhibition of DNA synthesis in combination with increased availability of ssDNA induces the SOS response resulting in up to 100-fold amplification of RecA protein and increased production of several other repair enzymes (3.1.2). As discussed in detail before (3.2), inhibition of DNA synthesis could cause an increase in the intracellular ATP and Mg^{2+} concentrations, stabilizing nascent RecA filaments and thus promoting RecA-mediated synapsis. However, amplified amounts of RecA and increased affinity of RecA toward DNA have a drawback — in the SOS-induced conditions, RecA is probably reluctant to dissociate from DNA after synapsis is achieved. Indeed, in vitro experiments show that at 10 mM Mg^{2+} RecA filaments do not disassemble from the products of DNA strand exchange provided that the products lack free ends, which corresponds to the situation with recombinational repair in vivo.[103-105] To remove used RecA filaments from recombining DNA under the SOS-induced conditions, another SOS-inducible activity may be needed, and RuvAB complex was suggested to have such an activity.[43,106] One of the predictions of this suggestion is that in *ruvAB* mutants, SOS induction in response to DNA damage should last longer than in WT cells,[106] which seems to be the case (unpublished observations cited in ref. 107). Also, there are *recA* mutants that partially suppress UV-sensitivity caused by deficiency in early proteins, RecF, RecO or RecR (ref. 16 and references therein). This suppression is thought to be due to increased rate of association of the mutant RecA proteins with ssDNA which somehow enhances the ability of RecA to displace SSB protein from ssDNA.[108,109] Interestingly, these *recA* mutations exacerbate UV sensitivity of *ruv* mutants,[13] as if Ruv proteins normally counteract RecA filament assembly. In fact, one of the first discovered phenotypes of *ruv* mutants was their incompatibility with the hyper-active RecA441 mutant protein.[73] However, the strongest indications that Holliday junctions can be used by RuvAB to dissociate used RecA filaments are: (1) the in vitro ability of RuvAB to dissociate recombinational intermediates covered by RecA[110,111] and (2) the in vitro ability of RuvAB to depolymerize RecA filament assembled on double-stranded DNA.[112] The latter reaction requires ATP hydrolysis and is observed only at Mg^{2+} concentrations 7.5 mM or higher.[112] The SOS-inducibility of *ruvA* and *ruvB* gene expression (4.5.1) is also consistent with the proposed role of RuvAB in removing used RecA filaments. There is another DNA helicase, Dda protein of the coliphage T4, which is argued to perform an analogous function in phage metabolism.[106]

EM photographs show that at certain RuvB concentrations, two hexameric rings of RuvB interact with junction-bound RuvA tetramer, sitting on the opposite sides of it (Fig. 4.6E), which is in a nice agreement with the current ideas.[100] However, the authors also report that at high concentrations, RuvB binds to all four arms of the junction. So, RuvA does not seem to direct RuvB binding to particular arms. What does? In vivo, Holliday junctions are likely to be always associated with RecA. RecA filament could preclude binding of RuvB to a particular pair of arms. Interestingly, the directionality of RuvB-promoted branch-migration[101] dictates that a Holliday junction will be pushed always towards the RecA filament, which may cause filament dissociation.

It is proposed that to favor RecA filament assembly, it is necessary to neutralize the highly negatively charged C-terminal domain of RecA protein (3.3). In normal conditions this neutralization could be achieved by interaction with the highly positively charged C-terminal domain of RecO protein (4.2.4). Conversely, to favor RecA filament disassembly, it should be sufficient to recharge the C-terminal domains of RecA monomers, and they will sever the filament. Under normal conditions, this recharging might be simply a result of joint molecule formation, when RecA filament has to hold three DNA strands instead of one. Under the SOS-induced conditions, when the negative charges of DNA are probably shielded by high concentrations of Mg^{2+} and spermidine, this recharging is likely to be mediated by a negatively-charged protein. Examination of the RuvA and RuvB protein sequences[77,78] reveals that RuvB, but not RuvA, is negatively-charged, which is consistent with its suggested facilitation of RecA filament disassembly.

4.5.5. RuvC is a Site-Specific Holliday Junction Resolvase

Holliday junctions with homologous cores, whether assembled of four oligonucleotides or formed in RecA-mediated recombinational reaction, are symmetrically incised in two opposite strands ("resolved") in *E. coli* crude extracts[113,114] (Fig. 4.5E). The cleaving activity was purified and ascribed to the product of *ruvC* gene.[81,115] RuvC protein can resolve not only deproteinized intermediates of an in vitro RecA-promoted strand exchange reaction, but also the Holliday junctions on DNAs still participating in a RecA-mediated reaction.[116]

In vitro, RuvC binds to Holliday junctions with an affinity 10^3-10^4 higher than when binding to linear duplex DNA.[117,118] The RuvC binding to Holliday junctions is observed at low temperature (0°C), is independent of divalent cations and is inhibited at Mg^{2+} concentrations over 1 mM with almost no binding seen at 10 mM Mg^{2+}. On the contrary, the RuvC cleavage of Holliday junctions requires high temperatures (37°C) and divalent cations with optimal Mg^{2+} concentration around 10 mM.[117-119] Cleavage occurs by nicking two strands of the same polarity. Nicks are placed the same distance from a presumed crossover junction. The nicking is site-specific — it occurs efficiently only in the sequence 5'-A/T-T-T-↓-G/C-3'.[120] Low level of nicking is also observed at degenerate consensus sequences having a single thymine residue on the 5' side of the cleavage site. RuvC binds to, but does not cleave, junctions that lack the nicking sequence.[120] Nicks introduced by RuvC can be directly sealed by *E. coli* DNA ligase.[115,117]

RuvC binds and resolves three-strand junctions as well, cutting a single, nonbranching strand. However, in contrast to the resolution of Holliday junctions, in vitro resolution of three-strand junctions by RuvC is observed only when RecA protein is completely removed.[121] Phage-encoded resolvases cleave a broad range of branched or irregular DNA structures[122] and could be considered "debranching" enzymes. In view of RuvC cleavage of structures other than orthodox four-strand junctions, the ability of RuvC to act on various irregular DNA structures was also verified. It was found that RuvC does not cleave Y-junctions, mismatches or heterologous loops.[118,121] Thus, in contrast to the phage debranching enzymes, RuvC seems to be strictly a recombinational repair resolvase.

4.5.6. Crystal Structure of RuvC. Possible Roles of the C-terminal Domain

The atomic structure of RuvC resolvase reveals a dimer formed by two wedge-shaped subunits (Fig. 4.7A) with two valleys on one side of the dimer[123] (Fig. 4.7B). The sides of these valleys are charged positively (Fig. 4.7C), which suggests that they interact with the negatively charged DNA backbone. A Holliday junction is proposed to undergo some isomerization (opening) before fitting into the two valleys of the RuvC dimer (Fig. 4.7B). Studies of RuvC mutants that are able to bind Holliday junctions, but are unable to resolve them, suggest that catalytic domains in the RuvC dimer are situated at the bottoms of the valleys and consist of four closely spaced negatively-charged residues[123,124] (Fig. 4.7A-C). The DNA strands that are proposed to lie across the active center (and so are likely to be cut during resolution) are the noncrossover strands (Fig. 4.7B). RuvC cleavage of the noncrossover strands was demonstrated with model Holliday junctions, whose branch-migration was restricted because one pair of arms was connected.[125]

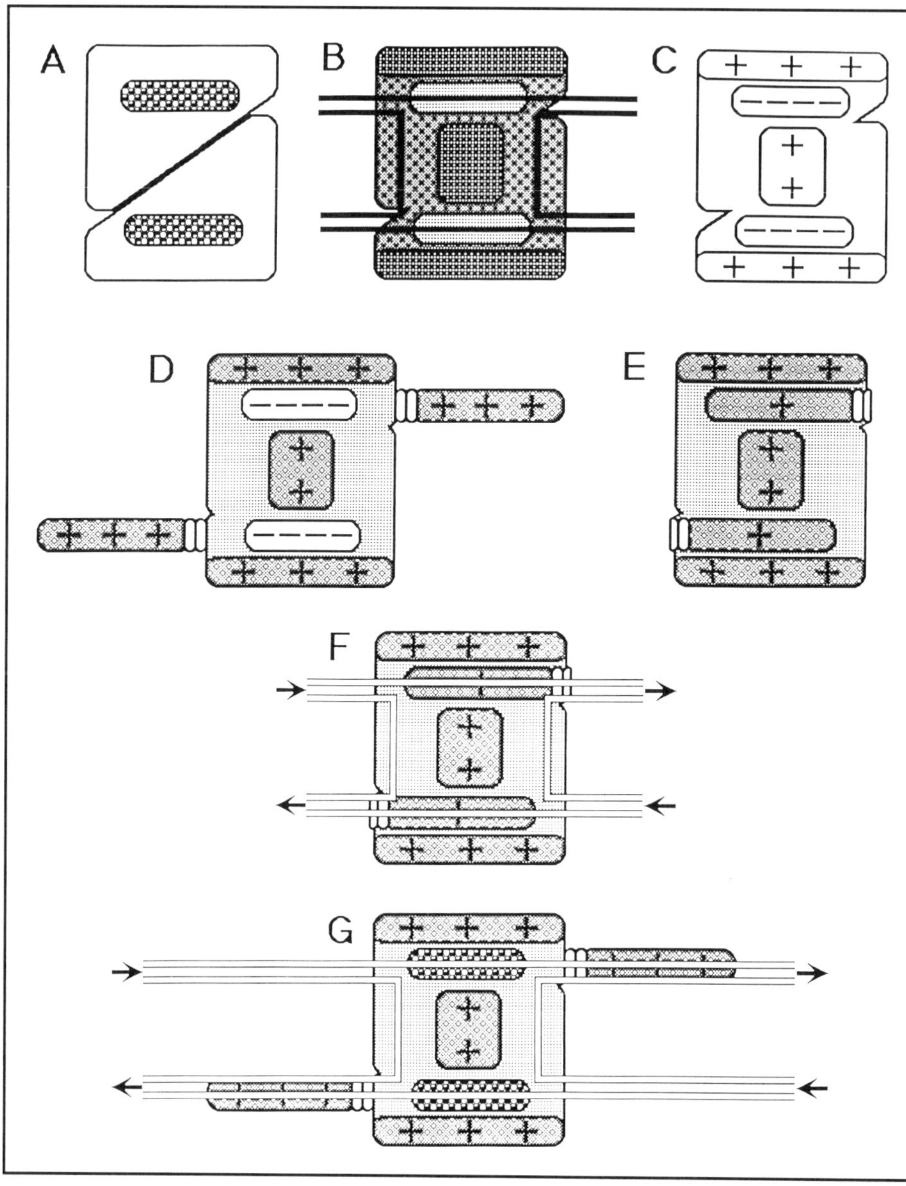

Fig. 4.7. A scheme of RuvC dimer, based on its atomic structure. (A) An overall shape of the dimer as seen from above, with two catalytic sites marked with bubbles. (B) An overall topography of the dimer as seen from above, with features above the plane of the paper in darker shades and with those below the plane of the paper in lighter shades. A Holliday junction in an open configuration is fitted onto the dimer. (C) A distribution of charges on the active surface of the RuvC dimer, as seen from above. Note the negatively-charged catalytic sites. (D) The same as in "C" but with the carboxy-terminal domains shown as the two positively charged belts. (E) The same as in "D", but the carboxy-terminal domains are shown to interact with the oppositely charged catalytic centers, rendering the enzyme inactive. (F) A Holliday junction is fitted onto the inactive enzyme shown in "E". (G) Branch migration of the Holliday junction shown by the arrows could "drag" the C-terminal domains out, unraveling the catalytic sites.

An interesting feature of the RuvC dimer is its positively charged 14-aminoacid C-terminal domain, which is disordered in crystals and may form a flexible tail[123] (Fig. 4.7D). In crystals, the acidic catalytic centers of RuvC molecules are partly blocked by the basic C-terminal tails belonging to other RuvC molecules. Since the tail is in close proximity to the acidic catalytic site of the same RuvC monomer and has the opposite charge, these two domains might interact with each other in vivo (Fig. 4.7E). Such interaction would make the DNA-binding valleys of a RuvC monomer positively-charged, facilitating DNA binding (Fig. 4.7F). On the other hand, catalytic centers of the resolvase would be masked by this domain interaction, rendering RuvC inactive in Holliday junction resolution. The enzyme could be activated for cutting only if the catalytic centers are unraveled as a result of the C-terminal domain dissociation. The unraveling could be achieved by branch migration of the junction — since DNA is negatively-charged, it should be able to "stick" to the basic tails and "drag" them out of the center of the RuvC dimer (Fig. 4.7G). This would explain a suspected in vivo requirement for branch-migration by RuvAB to ensure efficient resolution of Holliday junctions by RuvC.

The proposed auto-inhibition of RuvC should be relieved at alkaline pH which would cause the basic tail to lose its charges. Indeed, the optimal pH for RuvC-promoted resolution of Holliday junctions is 9.0.[119] Remarkably, a low degree of RuvC cutting of Holliday junctions lacking both the resolution consensus and homology does occur in the presence of Mn^{2+}, or at elevated temperatures, or at high pH.[126] These are all conditions favoring the uncovering of the catalytic center of the enzyme, and this cutting could be explained by a fraction of preactivated enzyme.

An alternative role for the C-terminal domain of RuvC might be interaction with RuvAB enzyme (4.5.7). The two ideas about possible roles of the C-terminal domain could be discriminated using truncated RuvC proteins lacking the domain. If the C-terminal domain of RuvC is important only for the interaction with RuvAB, a truncated resolvase might behave similarly to the WT enzyme in vitro, but be inactive in vivo. If, on the other hand, the C-terminal domain modulates the resolution activity of RuvC, a truncated protein might be permanently activated for cutting. Such an enzyme should be hyperactive in vitro but might be poisonous in vivo.

4.5.7. RuvC Unfolds Holliday Junctions. RuvAB + RuvC Resolvasome?

Although RuvC was crystallized without DNA, the determined atomic structure of the enzyme permitted a model Holliday junction to be fitted into its active center. Such modeling shows that the junction must be unstacked and partially unwound; this unfolded configuration is required because of the isle in the center of the RuvC dimer.

The model of RuvC binding to the junction was verified with the same method that was used to probe for the conformation of the junction bound by RuvA. It was found that the binding of RuvC, like that of RuvA, changes the conformation of the junction, unfolding it.[127]

Similar junction isomerization upon binding of RuvA or RuvC provokes intriguing questions. Do RuvA and RuvC compete for a junction? Could they bind a junction simultaneously? Could they work in concerted fashion or even in a single protein complex? The answer to these questions may lie in an asymmetry of Holliday junctions. A Holliday junction in the square planar conformation seems to be completely symmetrical, but in fact it has two distinct sides. On one side, DNA strands, if traced in the 5'-to-3' direction, go around the junction clockwise (Fig. 4.8A), while on the other side (imagine the junction flipped) the same strands go counter-clockwise (Fig. 4.8B). It is possible that both RuvA and RuvC bind Holliday junctions without respect to this polarity. However, it is an attractive possibility that these two activities bind to the opposite sides of the junctions. Addition of RuvB to this complex would complete the formation of a "resolvasome", a protein complex able to branch-migrate Holliday junctions to displace used RecA filament from DNA and to resolve the junction when a strong resolution site is encountered.

4.6. A BACK-UP REPAIR OF DAUGHTER-STRAND GAPS: TRANSLESION DNA SYNTHESIS

The recombinational repair of a daughter-strand gap is accomplished via RecA-promoted pairing of the gapped DNA with its intact sister duplex. However, when DNA is heavily damaged, recombinational repair of a daughter-strand gap might become unfeasible. This would hap-

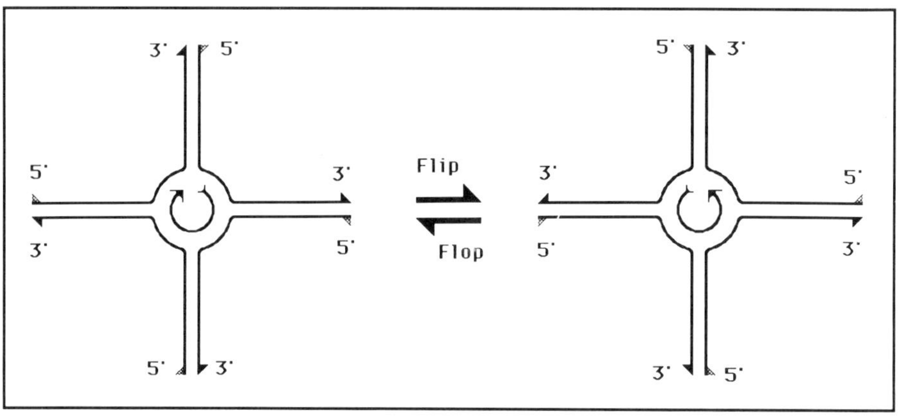

Fig. 4.8. The "flip" and the "flop" sides of a Holliday junction in an open planar configuration. 3' and 5' ends of DNA strands are marked. In the center, the 5'-to-3' directions around the junctions are shown with arrows. This is the same junction, but viewed from the opposite sides.

pen, for example, if both sister chromatids are damaged in the same sequence, or if the sister chromatid is degraded due to a downstream lesion. Many bacteria have a backup mechanism to repair lingering gaps without recombination, by translesion DNA synthesis. It is thought that RecA filament is disassembled from the gap, while the replisome is modified so as to be able to ignore the lesion. Interestingly, although nonrecombinational by its nature, this process is both regulated by RecA and requires RecA's direct participation (reviewed in refs. 128,129).

RecA regulates translesion DNA synthesis at two levels. The first level of regulation is via LexA cleavage and SOS induction. Prolonged SOS induction increases the expression of *umuDC* operon[130,131] (3.1); both UmuC and UmuD proteins are directly required for the translesion DNA synthesis. Before these two proteins can catalyze their reaction, UmuD needs a modification — RecA-promoted autocleavage of UmuD is the second level of RecA involvement in this process. UmuD cleaves itself in a reaction similar to the autocleavage of LexA and phage repressors.[132] In vitro this inefficient reaction, as well as autocleavage of phage repressors, requires higher concentrations of Mg^{2+} than RecA-promoted autocleavage of LexA; processing of UmuD in vivo is also rather inefficient.[133] This is likely to ensure that UmuD and phage repressors are cleaved late during SOS induction (3.1.3). As a result of the autocleavage, UmuD' protein is generated, which combines with UmuC to form a complex active in catalyzing translesion DNA synthesis.[134] UmuD' is a small negatively charged protein (a surplus of seven negatively–charged amino acids in a less than 120 amino acid polypeptide). In contrast, UmuC has a considerable positive charge (a surplus of 10 positively-charged amino acids in a 420 amino acid polypeptide).[135,136]

Besides its two regulatory roles in translesion DNA synthesis, RecA is required directly, together with UmuD', UmuC proteins and DNA pol III, for the translesion DNA synthesis to happen.[134] The current model envisions this process as UmuD'C-assisted reloading of a replisome at the 3'-end of the gap, which is still mostly covered by RecA filament (Fig. 4.9). Devoret and colleagues proposed[137] that UmuD'C complex dissociates RecA filament on one side while helping the replisome to polymerize through the noncoding lesion on the other side. This specific model is based on their observation that the expression of UmuD' and UmuC inhibits recombinational repair by interfering with RecA action. We would like to further detail their model using the facts that translesion DNA synthesis is inhibited by overproduction of DnaN "clamp" subunit of DNA pol III,[138] that overexpression of UmuC is poisonous for strains with defective DNA pol III[139] and that UmuD' is highly negatively charged (see above). We postulate that, by analogy with RuvB (4.5.4), UmuD' destabilizes RecA filaments, using its negative charges, while UmuC binds to DnaN clamp and modifies the loaded

DNA pol III, enabling translesion DNA synthesis (Fig. 4.9). Formation of the complex between UmuC and a DnaN clamp may be assisted by chaperones (matchmakers?), as discussed in ref.128. In this scheme, the direct role of RecA is in targeting UmuC protein via UmuD' protein to the particular DnaN clamp awaiting at the lesion. The prediction is that UmuC protein alone in sufficiently high concentrations will allow some translesion DNA synthesis by DNA pol III, if the latter is added after DnaN and UmuC (+ chaperones).

Since the replisome has to insert a random base opposite the noncoding lesion (which might be an abasic site), the translesion DNA synthesis often generates point mutations and is frequently referred to as induced mutagenesis. This is definitely not the best way to repair a lesion, but it is the cell's last resort (better to be mutated than dead), and *umu* mutations do confer moderate UV-sensitivity.[140] Besides, the translesion DNA synthesis is the only way to repair a lesion in a situation like conjugal transfer, when a single strand of plasmid DNA is transferred from one bacterial cell to another. During the following synthesis of the complementary strand there is no sister duplex to rely on in the case of a noncoding lesion in the template strand. In fact, many conjugative broad-host-range plasmids carry their own genes for translesion DNA synthesis.[141] This is helpful because even among the enterobacteria not all the species are capable of induced mutagenesis, even though almost all of them carry genes homologous to the *umuDC* operon of *E. coli.*[133]

SUMMARY

Daughter-strand gaps are formed when DNA replication encounters a noncoding lesion in template DNA and reinitiates downstream, leaving a single-stranded region adjacent to the lesion. The main mechanism for repair of daughter-strand gaps in *E. coli* is via synapsis of the gapped chromatid with the intact sister chromatid. The current scheme of this process is as follows[142] (Fig. 4.10): in preparation for the synapses, RecO and RecR proteins descend on the SSB-covered daughter-strand gap, targeted by RecF protein. The presence of RecO allows RecA polymerization on SSB-covered single-stranded DNA. In the synaptic phase of the reaction, RecA filament finds an intact duplex homologous to the single-strand gap and pairs them. The synapsis is facilitated by two distinct topoisomerases: DNA gyrase relieves positive supercoils generated in the intact duplex while topo I relieves negative supercoils, creating a new duplex between the invading and the resident strands. Pairing of the damaged and the intact DNA molecules allows filling-in of the gap by DNA polymerase I and repair of the lesion. In the postsynaptic phase of the reaction, RuvAB helicase disperses used RecA filaments and assists in RuvC-mediated resolution of Holliday junctions, completing faithful repair of the daughter-strand gap.

If an intact homologous duplex cannot be found, UmuD'C complex disperses RecA filament and modifies a replisome so as to allow translesion DNA synthesis. In this way, an intact DNA duplex is restored at the price of possible mutagenesis.

Other bacteria seem to mend daughter-strand gaps in their DNA using their RecF-analogous pathways, judging by the universal presence of

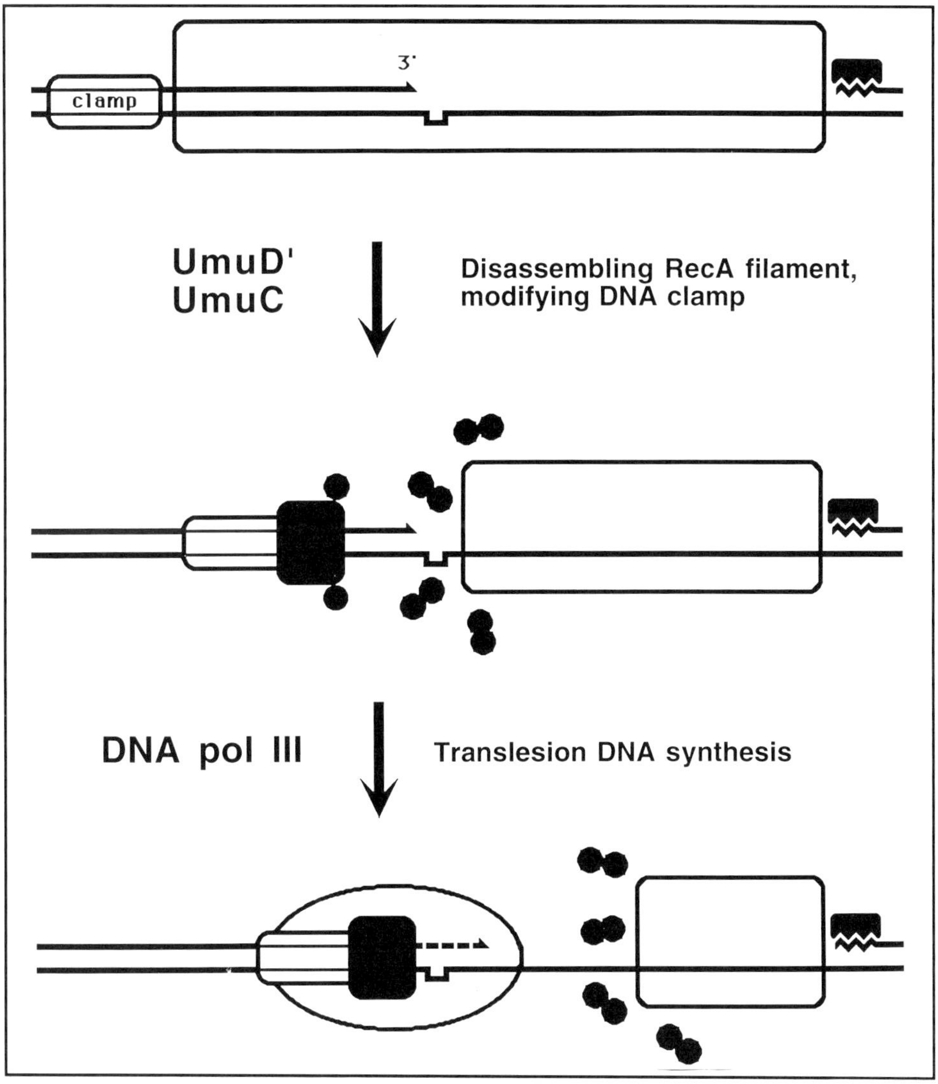

Fig. 4.9. A model for translesion DNA synthesis. RecA filament is shown as a big open rectangle; DnaN clamp is shown as a small open rectangle; UmuC is shown as a small black rectangle associated with DnaN clamp; UmuD is shown as small black circles, most of them dimers, some of them associated with UmuC; replisome is shown as an open oval. Explanations in the picture and in the text.

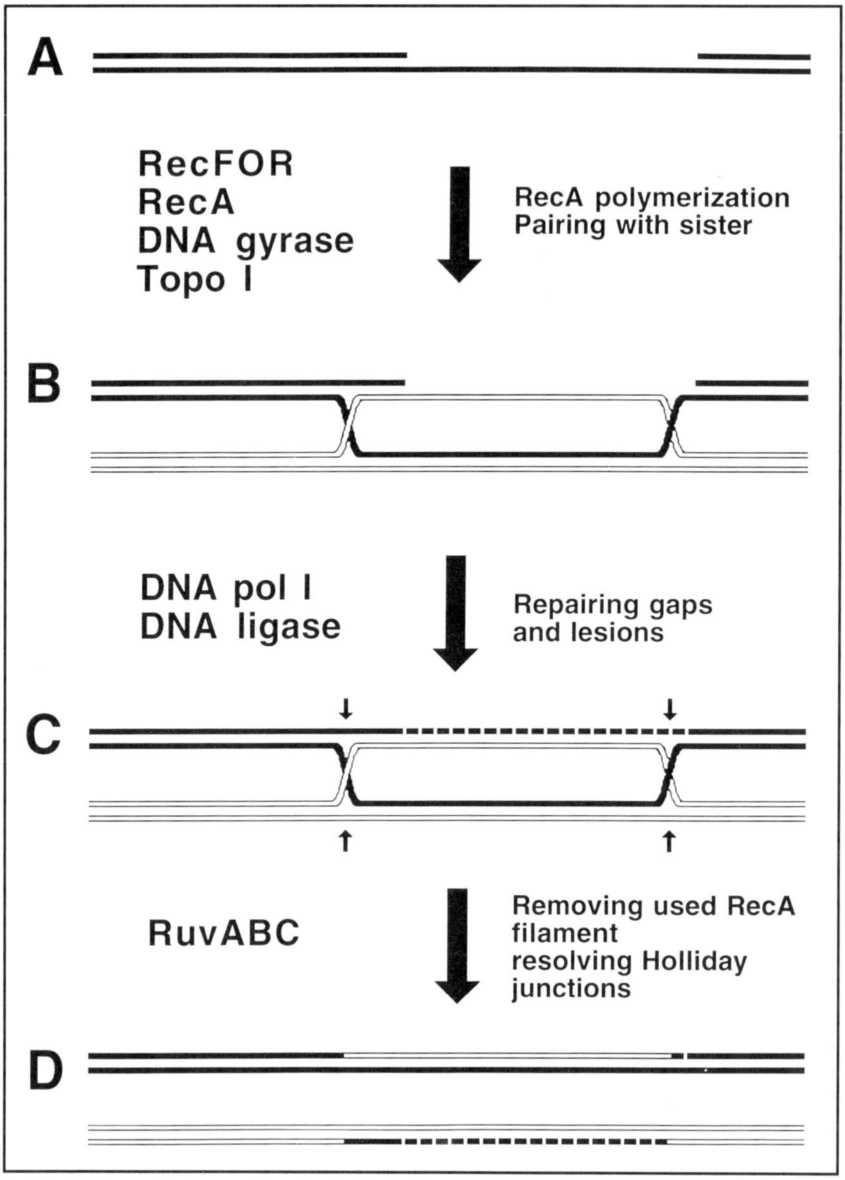

Fig. 4.10. An overview of daughter-strand gap repair. (A) A DNA molecule containing daughter-strand gap. The lesion that caused the gap is not shown. (B) Synapsis of the gapped molecule with the intact homologous DNA. (C) Repair of the gap and the lesion that caused the gap. (D) Disengaging the two DNA molecules. The RuvC cleavages are indicated by small vertical arrows in "C". Additional explanations in the text and in the figure itself.

recF homologs.[18,19] A homolog of *recR* is found in *B. subtilis*,[31] while *Haemophilus influenzae* has corresponding homologs for all the members of the *E. coli* RecF pathway.[143] Genes for translesion DNA synthesis are widespread among bacteria.[128,141]

Although the general scheme of daughter-strand gap repair in *E. coli* looks satisfying, the details of interaction of particular proteins are still poorly understood. For example, the exact mechanism of translesion DNA synthesis, and especially the intimate interaction of RecA filament with UmuD'C and replisome are yet to be elucidated.

The role of RecF in RecOR-catalyzed overcoming of SSB inhibition by RecA remains to be defined. It is proposed here that RecF binds to the 5'-side of every primer and is displaced by completion of the previous Okazaki piece. The mechanism of RecOR catalysis of RecA filament assembly on the SSB-covered ssDNA is also unclear. It is proposed here that positively-charged domains of RecO play a major role in stimulating RecA filament assembly.

The indistinguishable phenotypes of *ruvA*, *ruvB* and *ruvC* mutants hint to the possibility of direct interaction between RuvC resolvase and RuvAB complex in vivo. However, alternative explanations abound. For example, since RuvC cleaves Holliday junctions at specific sites in vitro, translocation of a Holliday junction by RuvAB might be necessary to position the junction at the RuvC cleavage site in vivo[120] (RuvAB-mediated enhancement of Holliday junction resolution by RuvC protein in vitro is yet to be demonstrated).

In vitro, RuvC can cleave Holliday junctions which are (at least partially) still within RecA filament.[116] However, in vivo, Holliday junctions within RecA filament might be a poor substrate for RuvC. Hence, another explanation for the similarity of *ruv* mutant phenotypes is that RuvAB complex has to remove RecA from Holliday junctions for it to be accessible to RuvC.

Still another explanation for the similarity of *ruv* mutant phenotypes attributes it to related, although mechanistically different reasons. It assumes that in *ruv* mutants recombining chromosomes are joined together — in *ruvAB* mutants they are held together by RecA filaments, in *ruvC* mutants, by Holliday junctions. In vitro studies with all three proteins added simultaneously to RecA-covered joint molecules should be able to distinguish between these possibilities. The attractive idea that RuvA, RuvB and RuvC proteins act in a single complex, resolvasome, calls for experimental testing.

References

1. Rupp WD and Howard-Flanders P (1968) Discontinuities in the DNA synthesized in an excision-defective strain of *Escherichia coli* following ultraviolet irradiation. J Mol Biol 31:291-304.
2. Iyer VN and Rupp WD (1971) Usefulness of benzoylated naphthoylated DEAE-cellulose to distinguish and fractionate double-stranded DNA bearing different extents of single-stranded regions. Biochim Biophys Acta 228:117-126.

3. Kornberg A and Baker TA (1992) DNA Replication. New York: W.H. Freeman and Company, 931 p.
4. Wang T-CV and Chen S-H (1992) Similar-sized daughter-strand gaps are produced in the leading and the lagging strands of DNA in UV-irradiated *E. coli uvrA* cells. Biochem Biophys Res Com 184:1496-1503.
5. Rupp WD, Wilde III CE, Reno DL and Howard-Flanders P (1971) Exchanges between DNA strands in ultraviolet-irradiated *Escherichia coli.* J Mol Biol 61:25-44.
6. Howard-Flanders P (1973) DNA repair and recombination. Brit Med Bull 29:226-235.
7. Smith KC and Meun DHC (1970) Repair of radiation-induced damage in *Escherichia coli.* I. Effect of *rec* mutations on post-replication repair of damage due to ultraviolet radiation. J Mol Biol 51:459-472.
8. Ganesan AK (1974) Persistence of pyrimidine dimers during post-replication repair in ultraviolet light-irradiated *Escherichia coli* K12. J Mol Biol 87:103-119.
9. Horii ZI and Clark AJ (1973) Genetic analysis of the RecF pathway of genetic recombination in *Escherichia coli* K-12: isolation and characterization of mutants. J Mol Biol 80:327-344.
10. Rothman RH and Kato T (1975) The beginning of an investigation of the role of *recF* in the pathways of metabolism of ultraviolet-irradiated DNA in *Escherichia coli.* In: Hanawalt PC and Setlow RB, eds. Molecular Mechanisms for Repair of DNA. New York: Plenum Press, pp 283-291.
11. Lloyd RG, Porton MC and Buckman C (1988) Effect of *recF, recJ, recN, recO* and *ruv* mutations on ultraviolet survival and genetic recombination in a *recD* strain of *Escherichia coli* K12. Mol Gen Genet 212:317-324.
12. Mahdi AA and Lloyd RG (1989) Identification of the *recR* locus of *Escherichia coli* K-12 and analysis of its role in recombination and DNA repair. Mol Gen Genet 216:503-510.
13. Tseng Y-C, Hung J-L and Wang T-CV (1994) Involvement of the RecF pathway recombination genes in postreplication repair in UV-irradiated *Escherichia coli* cells. Mutat Res 315:1-9.
14. Volkert MR and Hartke MA (1984) Suppression of *Escherichia coli recF* mutations by *recA*-linked *srfA* mutations. J Bacteriol 169:498-506.
15. Volkert MR, Margossian LJ and Clark AJ (1984) Two-component suppression of *recF143* by *recA441* in *Escherichia coli* K-12. J Bacteriol 160:702-705.
16. Wang T-CV, Chang H-Y and Hung J-L (1993) Cosuppression of *recF, recR* and *recO* mutations by mutant *recA* alleles in *Escherichia coli* cells. Mutat Res 294:157-166.
17. Perez-Roger I, Garcia-Sogo M, Navarro-Avino JP, Lopez-Acedo C, Macian F and Armengod ME (1991) Positive and negative regulatory elements in the *dnaA-dnaN-recF* operon of *Escherichia coli.* Biochimie 73:329-334.
18. Fujita MQ, Yoshikawa H and Ogasawara N (1989) Structure of *dnaA* region of *Pseudomonas putida*: conservation among three bacteria, *Bacillus subtilis, Escherichia coli* and *P. putida.* Mol Gen Genet 215:381-387.

19. Margerrison EEC, Hopewell R and Fisher LM (1992) Nucleotide sequence of the *Staphylococcus aureus gyrB-gyrA* locus encoding the DNA gyrase A and B proteins. J Bacteriol 174:1596-1603.
20. Sandler SJ and Clark AJ (1990) Factors affecting expression of the *recF* gene of *Escherichia coli* K-12. Gene 86:35-43.
21. Armengod ME, Garcia-Sogo M, Perez-Roger I, Macian F and Navarro-Avino JP (1991) Tandem transcription termination sites in the *dnaN* gene of *Escherichia coli.* J Biol Chem 266:19725-19730.
22. Madiraju MVVS and Clark AJ (1991) Effect of RecF protein on reactions catalyzed by RecA protein. Nucleic Acids Res 19:6295-6300.
23. Burgers PMJ, Kornberg A and Sakakibara Y (1981) The *dnaN* gene codes for the β subunit of DNA polymerase III holoenzyme of *Escherichia coli.* Proc Natl Acad Sci USA 78:5391-5395.
24. Griffin IV TJ and Kolodner RD (1990) Purification and preliminary characterization of the *Escherichia coli* K-12 RecF protein. J Bacteriol 172:6291-6299.
25. Sandler SJ and Clark AJ (1993) Use of high and low level overexpression plasmids to test mutant alleles of the *recF* gene of *Escherichia coli* K-12 for partial activity. Genetics 135:643-654.
26. Morrison PT, Lovett ST, Gilson LE and Kolodner R (1989) Molecular analysis of the *Escherichia coli recO* gene. J Bacteriol 171:3641-3649.
27. Takiff HE, Chen S-M and Court DL (1989) Genetic analysis of the *rnc* operon of *Escherichia coli.* J Bacteriol 171:2581-2590.
28. Lin Y-P, Sharer JD and March PE (1994) GTPase-dependent signaling in bacteria: characterization of a membrane-binding site for Era in *Escherichia coli.* J Bacteriol 176:44-49.
29. Yeung T, Mullin DA, Chen K-S, Craig EA, Bardwell JCA and Walker JR (1990) Sequence and expression of the *Escherichia coli recR* locus. J Bacteriol 172:6042-6047.
30. Flower AM and McHenry CS (1991) Transcriptional organization of the *Escherichia coli dnaX* gene. J Mol Biol 220:649-658.
31. Alonso JC, Shirahige K and Ogasawara N (1990) Molecular cloning, genetic characterization and DNA sequence analysis of the *recM* region of *Bacillus subtilis.* Nucleic Acids Res 18:6771-6777.
32. Madiraju MVVS and Clark AJ (1992) Evidence for ATP binding and double-stranded DNA binding by *Escherichia coli* RecF protein. J Bacteriol 174:7705-7710.
33. Sandler SJ, Chackerian B, Li JT and Clark AJ (1992) Sequence and complementation analysis of *recF* genes from *Escherichia coli, Salmonella typhimurium, Pseudomonas putida* and *Bacillus subtilis*: evidence for an essential phosphate binding loop. Nucleic Acids Res 20:839-845.
34. Luisi-DeLuca C and Kolodner R (1994) Purification and characterization of the *Escherichia coli* RecO protein. Renaturation of complementary single-stranded DNA molecules catalyzed by the RecO protein. J Mol Biol 236:124-138.

35. Alonso JC, Stiege AC, Dobrinski B and Lurz R (1993) Purification and properties of the RecR protein from *Bacillus subtilis* 168. J Biol Chem 268:1424-1429.
36. Umezu K, Chi N-W and Kolodner RD (1993) Biochemical interaction of the *Escherichia coli* RecF, RecO and RecR proteins with RecA and single-stranded DNA binding protein. Proc Natl Acad Sci USA 90:3875-3879.
37. Umezu K and Kolodner RD (1994) Protein interactions in genetic recombination in *Escherichia coli*. Interactions involving RecO and RecR overcome the inhibition of RecA by single-stranded DNA-binding protein. J Biol Chem 269:30005-30013.
38. Yonesaki T and Minagawa T (1985) T4 phage gene *uvsX* product catalyzes homologous DNA pairing. EMBO J 4:3321-3327.
39. Yonesaki T and Minagawa T (1989) Synergistic action of three recombination gene products of bacteriophage T4, UvsX, UvsY, and gene 32 proteins. J Biol Chem 264:7814-7820.
40. Jiang H, Giedroc D and Kodadek T (1993) The role of protein-protein interactions in the assembly of the presynaptic filament for T4 homologous recombination. J Biol Chem 268:7904-7911.
41. Sandler SJ and Clark AJ (1994) RecOR suppression of *recF* mutant phenotypes in *Escherichia coli* K-12. J Bacteriol 176:3661-3672.
42. Mahdi AA and Lloyd RG (1989) The *recR* locus of *Escherichia coli* K-12: molecular cloning, DNA sequencing and identification of the gene product. Nucleic Acids Res 17:6781-6794.
43. Kuzminov A (1995) A mechanism for induction of the SOS response in *E. coli*: insights into the regulation of reversible protein polymerizaton in vivo. J Theor Biol 177:29-43.
44. Freitag N and McEntee K (1988) Affinity chromatography of RecA protein and RecA nucleoprotein complexes on RecA protein-agarose column. J Biol Chem 263:19525-19534.
45. Thoms B and Wackernagel W (1988) Suppression of the UV-sensitive phenotype of *Escherichia coli recF* mutants by *recA*(Srf) and *recA*(Tif) mutations requires *recJ*$^+$. J Bacteriol 170:3675-3681.
46. Clark AJ, Volkert MR, Margossian LJ and Nagaishi H (1982) Effects of a *recA* operator mutation on mutant phenotypes conferred by *lexA* and *recF* mutations. Mutat Res 106:11-26.
47. Thoms B and Wackernagel W (1987) Regulatory role of *recF* in the SOS response of *Escherichia coli*: impaired induction of SOS genes by UV irradiation and nalidixic acid in a *recF* mutant. J Bacteriol 169:1731-1736.
48. Kelner A (1953) Growth, respiration, and nucleic acid synthesis in ultraviolet-irradiated and in photoreactivated *Escherichia coli*. J Bacteriol 65:252-262.
49. Khidhir MA, Casaregola S and Holland IB (1985) Mechanism of transient inhibition of DNA synthesis in ultraviolet-irradiated *E. coli*: inhibition is independent of *recA* whilst recovery requires RecA protein itself and an additional, inducible SOS function. Mol Gen Genet 199:133-140.

50. Witkin EM, Roegner-Maniscalco V, Sweasy JB and McCall JO (1987) Recovery from ultraviolet light-induced inhibition of DNA synthesis requires *umuDC* gene products in *recA718* mutant strains but not in *recA*+ strains of *Escherichia coli*. Proc Natl Acad Sci USA 84:6805-6809.
51. Doudney CO (1990) DNA-replication recovery inhibition and subsequent reinitiation in UV-radiation-damaged *E. coli*: a strategy for survival. Mutat Res 243:179-186.
52. Altshuler M (1993) Recovery of DNA replication in UV-damaged *Escherichia coli*. Mutat Res 294:91-100.
53. Bremer H and Dennis PP (1987) Modulation of chemical composition and other parameters of the cell by growth rate. In: Neidhardt FC, ed. *Escherichia coli* and *Salmonella typhimurium*. Cellular and molecular biology. Washington, D.C.: American Society for Microbiology, pp 1527-1542.
54. Wu YH, Franden MA, Hawker JR and McHenry CS (1984) Monoclonal antibodies specific for the α subunit of the *Escherichia coli* DNA polymerase III holoenzyme. J Biol Chem 259:12117-12122.
55. Granger-Schnarr M, Schnarr M and van Sluis CA (1986) In vitro study of the interaction of the LexA repressor and the UvrC protein with a *uvrC* regulatory region. FEBS Lett 198:61-65.
56. Payne NS and Sancar A (1989) The LexA protein does not bind specifically to the two SOS box-like sequences immediately 5' to the *phr* gene. Mutat Res 218:207-210.
57. Quinones A, Jueterbock W-R and Messer W (1991) DNA lesions that block DNA replication are responsible for the *dnaA* induction caused by DNA damage. Mol Gen Genet 231:81-87.
58. Kaasch M, Kaasch J and Quinones A (1989) Expression of *dnaN* and *dnaQ* genes of *Escherichia coli* is inducible by mitomycin C. Mol Gen Genet 219:187-192.
59. Tadmor Y, Bergstein M, Skaliter R, Shwartz H and Livneh Z (1994) β subunit of DNA polymerase III holoenzyme is induced upon ultraviolet irradiation or nalidixic acid treatment of *Escherichia coli*. Mutat Res 308:53-64.
60. Burgers PMJ and Kornberg A (1983) The cycling of *Escherichia coli* DNA polymerase III holoenzyme in replication. J Biol Chem 258:7669-7675.
61. Shavitt O and Livneh Z (1989) The β subunit modulates bypass and termination at UV-lesions during in vitro replication with DNA polymerase III holoenzyme of *Escherichia coli*. J Biol Chem 264:11275-11281.
62. O'Donnell M and Studwell PS (1990) Total reconstitution of DNA polymerase III holoenzyme from *Escherichia coli*. J Biol Chem 265:1179-1187.
63. Hays JB and Boehmer S (1978) Antagonists of DNA gyrase inhibit repair and recombination of UV-irradiated phage λ. Proc Natl Acad Sci USA 75:4125-4129.
64. von Wright A and Bridges BA (1981) Effect of *gyrB*-mediated changes in chromosome structure on killing of *Escherichia coli* by ultraviolet light: experiments with strains differing in deoxyribonucleic acid repair capacity. J Bacteriol 146:18-23.

65. Reece RJ and Maxwell A (1991) DNA gyrase: structure and function. Crit Rev Biochem Mol Biol 26:335-375.
66. Overbye KM and Margolin P (1981) Role of the *supX* gene in ultraviolet light-induced mutagenesis in *Salmonella typhimurium*. J Bacteriol 146:170-178.
67. Sternglanz R, DiNardo S, Voelkel KA, Nishimura Y, Hirota Y, Becherer K, Zumstein L and Wang JC (1981) Mutations in the gene coding for *Escherichia coli* DNA topoisomerase I affect transcription and transposition. Proc Natl Acad Sci USA 78:2747-2751.
68. Trucksis M and Depew RE (1981) Identification and localization of a gene that specifies production of *Escherichia coli* DNA topoisomerase I. Proc Natl Acad Sci USA 78:2164-2168.
69. Holliday R (1964) A mechanism for gene conversion in fungi. Genet Res 5:282-304.
70. Cooper PK (1982) Characterization of long patch excision repair of DNA in ultraviolet-irradiated *Escherichia coli*: an inducible function under Rec-Lex control. Mol Gen Genet 185:189-197.
71. Otsuji N, Iyehara H and Hideshima Y (1974) Isolation and characterization of *Escherichia coli ruv* mutant which forms nonseptate filaments after low doses of ultraviolet light irradiation. J Bacteriol 117:337-344.
72. Lloyd RG, Benson FE and Shurvinton CE (1984) Effect of *ruv* mutations on recombination and DNA repair in *Escherichia coli* K12. Mol Gen Genet 194:303-309.
73. Otsuji N and Iyehara-Ogawa H (1979) Thermoresistant revertants of an *Escherichia coli* strain carrying *tif-1* and *ruv* mutations: non-suppressibility of *ruv* by *sfi*. J Bacteriol 138:1-6.
74. Lloyd RG (1991) Conjugational recombination in resolvase-deficient *ruvC* mutants of *Escherichia coli* K-12 depends on *recG*. J Bacteriol 173: 5414-5418.
75. Benson F, Collier S and Lloyd RG (1991) Evidence of abortive recombination in *ruv* mutants of *Escherichia coli* K12. Mol Gen Genet 225:266-272.
76. Shurvinton CE and Lloyd RG (1982) Damage to DNA induces expression of the *ruv* gene of *Escherichia coli*. Mol Gen Genet 185:352-355.
77. Benson FE, Illing GT, Sharples GJ and Lloyd RG (1988) Nucleotide sequencing of the *ruv* region of *Escherichia coli* K12 reveals a LexA regulated operon encoding two genes. Nucleic Acids Res 16:1541-1549.
78. Shinagawa H, Makino K, Amemura M, Kimura S, Iwasaki H and Nakata A (1988) Structure and regulation of the *Escherichia coli ruv* operon involved in DNA repair and recombination. J Bacteriol 170:4322-4329.
79. Sharples GJ, Benson FE, Illing GT and Lloyd RG (1990) Molecular and functional analysis of the *ruv* region of *Escherichia coli* K-12 reveals three genes involved in DNA repair and recombination. Mol Gen Genet 221:219-226.
80. Evans DH and Kolodner R (1987) Construction of a synthetic Holliday junction analog and characterization of its interaction with a *Saccharomy-*

ces cerevisiae endonuclease that cleaves Holliday junctions. J Biol Chem 262:9160-9165.

81. Connolly B, Parsons CA, Benson FE, Dunderdale HJ, Sharples GJ, Lloyd RG and West SC (1991) Resolution of Holliday junctions in vitro requires the *Escherichia coli ruvC* gene product. Proc Natl Acad Sci USA 88:6063-6067.
82. Tsaneva IR, Illing G, Lloyd RG and West SC (1992) Purification and properties of the RuvA and RuvB proteins of *Escherichia coli*. Mol Gen Genet 235:1-10.
83. Shiba T, Iwasaki H, Nakata A and Shinagawa H (1993) *Escherichia coli* RuvA and RuvB proteins involved in recombinational repair: physical properties and interaction with DNA. Mol Gen Genet 237:395-399.
84. Shiba T, Iwasaki H, Nakata A and Shinagawa H (1991) SOS-inducible DNA repair proteins, RuvA and RuvB, of *Escherichia coli*: functional interaction between RuvA and RuvB for ATP hydrolysis and renaturation of the cruciform structure in supercoiled DNA. Proc Natl Acad Sci USA 88:8445-8449.
85. Muller B, Tsaneva IR and West SC (1993) Branch migration of Holliday junctions promoted by the *Escherichia coli* RuvA and RuvB proteins. II. Interaction of RuvB with DNA. J Biol Chem 268:17185-17189.
86. Iwasaki H, Shiba T, Nakata A and Shinagawa H (1989) Overproduction, purification and ATPase activity of the *Escherichia coli* RuvB protein involved in DNA repair. J Bacteriol 171:5276-5280.
87. Matson SW (1991) DNA helicases of *Escherichia coli*. Prog Nucl Acid Res Mol Biol 40:289-326.
88. Tsaneva IR, Muller B and S.C. W (1993) The RuvA and RuvB proteins of *Escherichia coli* exhibit DNA helicase activity in vitro. Proc Natl Acad Sci USA 90:1315-1319.
89. Stasiak A, Tsaneva IR, West SC, Benson CJB, Yu X and Egelman EH (1994) The *Escherichia coli* RuvB branch migration protein forms double hexameric rings around DNA. Proc Natl Acad Sci USA 91:7618-7622.
90. Iwasaki H, Takahagi M, Nakata A and Shinagawa H (1992) *Escherichia coli* RuvA and RuvB proteins specifically interact with Holliday junctions and promote branch migration. Genes Dev 6:2214-2220.
91. Parsons CA, Tsaneva I, Lloyd RG and West SC (1992) Interaction of *Escherichia coli* RuvA and RuvB proteins with synthetic Holliday junctions. Proc Natl Acad Sci USA 89:5452-5456.
92. Parsons CA and West SC (1993) Formation of a RuvAB-Holliday junction complex in vitro. J Mol Biol 232:397-405.
93. Menetski JP and Kowalczykowski SC (1989) Enhancement of *Escherichia coli* RecA protein enzymatic function by dATP. Biochemistry 28: 5871-5881.
94. Muller B, Tsaneva IR and West SC (1993) Branch migration of Holliday junctions promoted by the *Escherichia coli* RuvA and RuvB proteins. I. Comparison of RuvAB- and RuvB-mediated reactions. J Biol Chem 268:17179-17184.

95. Shibata T, DasGupta C, Cunningham RP, Williams JGK, Osber L and Radding CM (1981) Homologous pairing in genetic recombination. The pairing reaction catalyzed by *Escherichia coli* RecA protein. J Biol Chem 256:7565-7572.
96. Roman LJ and Kowalczykowski SC (1986) Relationship of the physical and enzymatic properties of *Escherichia coli* RecA protein to its strand exchange activity. Biochemistry 25:7375-7385.
97. Duckett DR, Murchie AIH, Diekmann S, von Kitzing E, Kemper B and Lilley DMJ (1988) The structure of the Holliday junction, and its resolution. Cell 55:79-89.
98. von Kitzing E, Lilley DMJ and Diekmann S (1990) The stereochemistry of a four-way DNA junction: a theoretical study. Nucleic Acids Res 18:2671-2683.
99. Panyutin IG, Biswas I and Hsieh P (1995) A pivotal role for the structure of the Holliday junction in DNA branch migration. EMBO J 14:1819-1826.
100. Parsons CA, Stasiak A, Bennett RJ and West SC (1995) Structure of a multisubunit complex that promotes DNA branch migration. Nature 374:375-378.
101. Hiom K and West SC (1995) Branch migration during homologous recombination: assembly of a RuvAB-Holliday junction complex in vitro. Cell 80:787-793.
102. Adams DE and West SC (1995) Unwinding of closed circular DNA by the *Escherichia coli* RuvA and RuvB recombination/repair proteins. J Mol Biol 247:404-417.
103. Pugh BF and Cox MM (1987) RecA protein binding to the heteroduplex products of DNA strand exchange. J Biol Chem 262:1337-1343.
104. Pugh BF and Cox MM (1987) Stable binding of RecA protein to duplex DNA. Unraveling a paradox. J Biol Chem 262:1326-1336.
105. Lindsley JE and Cox MM (1989) Dissociation pathway for RecA nucleoprotein filaments formed on linear duplex DNA. J Mol Biol 205:695-711.
106. Kuzminov A (1993) RuvA, RuvB and RuvC proteins: cleaning-up after recombinational repairs in *E. coli*. BioEssays 15:355-358.
107. Shinagawa H and Iwasaki H (1995) Molecular mechanisms of Holliday junction processing in *Escherichia coli*. Adv Biophys 31:49-65.
108. Madiraju MVVS, Templin A and Clark AJ (1988) Properties of a mutant *recA*-encoded protein reveal a possible role for *Escherichia coli recF*-encoded protein in genetic recombination. Proc Natl Acad Sci USA 85:6592-6596.
109. Madiraju MVVS, Lavery PE, Kowalczykowski SC and Clark AJ (1992) Enzymatic properties of the RecA803 protein, a partial suppressor of *recF* mutations. Biochemistry 31:10529-10535.
110. Tsaneva IR, Muller B and West SC (1992) ATP-dependent branch migration of Holliday junctions promoted by the RuvA and RuvB proteins of *E. coli*. Cell 69:1171-1180.

111. Iype LE, Inman RB and Cox MM (1995) Blocked RecA protein-mediated DNA strand exchange reactions are reversed by the RuvA and RuvB proteins. J Biol Chem 270:19473-19480.
112. Adams DE, Tsaneva IR and West SC (1994) Dissociation of RecA filaments from duplex DNA by the RuvA and RuvB DNA repair proteins. Proc Natl Acad Sci USA 91:9901-9905.
113. Conolly B and West SC (1990) Genetic recombination in *E. coli*: Holliday junctions made by RecA protein are resolved by fractionated cell-free extracts. Proc Natl Acad Sci USA 87:8476-8480.
114. Taylor AF and Smith GR (1990) Action of RecBCD enzyme on cruciform DNA. J Mol Biol 211:117-134.
115. Iwasaki H, Takahagi M, Shiba T, Nakata A and Shinagawa H (1991) *Escherichia coli* RuvC protein is an endonuclease that resolves the Holliday structure. EMBO J 10:4381-4389.
116. Dunderdale HJ, Benson FE, Parsons CA, Sharples GJ, Lloyd RG and West SC (1991) Formation and resolution of recombination intermediates by *E. coli* RecA and RuvC proteins. Nature 354:506-510.
117. Bennett RJ, Dunderdale HJ and West SC (1993) Resolution of Holliday junctions by RuvC resolvase: cleavage specificity and DNA distortion. Cell 74:1021-1031.
118. Takahagi M, Iwasaki H and Shinagawa H (1994) Structural requirements of substrate DNA for binding to and cleavage by RuvC, a Holliday junction resolvase. J Biol Chem 269:15132-15139.
119. Dunderdale HJ, Sharples GJ, Lloyd RG and West SC (1994) Cloning, overexpression, purification, and characterization of the *Escherichia coli* RuvC Holliday junction resolvase. J Biol Chem 269:5187-5194.
120. Shah R, Bennett RJ and West SC (1994) Genetic recombination in *E. coli*: RuvC protein cleaves Holliday junctions at resolution hotspots in vitro. Cell 79:853-864.
121. Benson FE and West SC (1994) Substrate specificity of the *Escherichia coli* RuvC protein. Resolution of three- and four-stranded recombination intermediates. J Biol Chem 269:5195-5201.
122. Mosig G (1994) Homologous recombination. In: Karam JD, ed. Molecular Biology of Bacteriophage T4. Washington, D.C.: American Society for Microbiology, pp 54-82.
123. Ariyoshi M, Vassylyev DG, Iwasaki H, Nakamura H, Shinagawa H and Morikawa K (1994) Atomic structure of the RuvC resolvase: a Holliday junction-specific endonuclease from *E. coli*. Cell 78:1063-1072.
124. Saito A, Iwasaki H, Ariyoshi M, Morikawa K and Shinagawa H (1995) Identification of four acidic amino acids that constitute the catalytic center of the RuvC Holliday junction resolvase. Proc Natl Acad Sci USA 92:7470-7474.
125. Bennett RJ and West SC (1995) RuvC protein resolves Holliday junctions via cleavage of the continuous (noncrossover) strands. Proc Natl Acad Sci USA 92:5635-5639.

126. Shah R, Bennett RJ and West SC (1994) Activation of RuvC Holliday junction resolvase in vitro. Nucleic Acids Res 22:2490-2497.
127. Bennett RJ and West SC (1995) Structural analysis of the RuvC-Holliday junction complex reveals an unfolded junction. J Mol Biol 252:213-226.
128. Woodgate R and Sedgwick SG (1992) Mutagenesis induced by bacterial UmuDC proteins and their plasmid homologues. Mol Microbiol 6:2213-2218.
129. Livneh Z, Cohen-Fix O, Skaliter R and Elizur T (1993) Replication of damaged DNA and the molecular mechanism of ultraviolet light mutagenesis. Crit Rev Biochem Mol Biol 28:465-513.
130. Bagg A, Kenyon CJ and Walker GC (1981) Inducibility of a gene product required for UV and chemical mutagenesis in *Escherichia coli*. Proc Natl Acad Sci USA 78:5749-5753.
131. Bailone A, Sommer S, Knezevic J, Dutreix M and Devoret R (1991) A RecA protein mutant deficeint in its interaction with the UmuDC complex. Biochimie 73:479-484.
132. Burckhardt SE, Woodgate R, Scheuermann RH and Echols H (1988) UmuD mutagenesis protein of *Escherichia coli*: overproduction, purification, and cleavage by RecA. Proc Natl Acad Sci USA 85:1811-1815.
133. Sedgwick SG, Ho C and Woodgate R (1991) Mutagenic DNA repair in enterobacteria. J Bacteriol 173:5604-5611.
134. Rajagopalan M, Lu C, Woodgate R, O'Donnell M, Goodman MF and Echols H (1992) Activity of the purified mutagenesis proteins UmuC, UmuD', and RecA in replicative bypass of an abasic DNA lesion by DNA polymerase III. Proc Natl Acad Sci USA 89:10777-10781.
135. Kitagawa Y, Akaboshi E, Shinagawa H, Horii T, Ogawa H and Kato T (1985) Structural analysis of the *umu* operon required for inducible mutagenesis in *Escherichia coli*. Proc Natl Acad Sci USA 82:4336-4340.
136. Perry KL, Elledge SJ, Mitchell BB, Marsh L and Walker GC (1985) *umuDC* and *mucAB* operons whose products are required for UV light- and chemical-induced mutagenesis: UmuD, MucA and LexA proteins share homology. Proc Natl Acad Sci USA 82:4331-4335.
137. Sommer S, Bailone A and Devoret R (1993) The appearance of the UmuD'C protein complex in *Escherichia coli* switches repair from homologous recombination to SOS mutagenesis. Mol Microbiol 10:963-971.
138. Tadmor Y, Ascarelli-Goell R, Skaliter R and Livneh Z (1992) Overproduction of the β subunit of DNA polymerase III holoenzyme reduces UV mutagenesis in *Escherichia coli*. J Bacteriol 174:2517-2524.
139. Nowicka A, Kanabus M, Sledziewska-Gojska E and Ciesla Z (1994) Different UmuC requirements for generation of different kinds of UV-induced mutations in *Escherichia coli*. Mol Gen Genet 243:584-592.
140. Kato T and Shinoura Y (1977) Isolation and characterization of mutants of *Escherichia coli* deficient in induction of mutations by ultraviolet light. Mol Gen Genet 156:121-131.

141. Strike P and Lodwick D (1987) Plasmid genes affecting DNA repair and mutation. J Cell Sci Suppl 6:303-321.
142. Clark AJ and Sandler SJ (1994) Homologous recombination: the pieces begin to fall into place. Crit Rev Microbiol 20:125-142.
143. Fleischmann RD, Adams MD, White O et al. (1995) Whole-genome random sequencing and assembly of *Haemophilus influenzae* Rd. Science 269:496-512.

CHAPTER 5

Double-Strand End Repair

5.1. THE ORIGIN OF DOUBLE-STRAND BREAKS AND DOUBLE-STRAND ENDS

When DNA molecules are centrifuged through a neutral sucrose gradient, their sedimentation rate is a function of their molecular weight: the longer molecules sediment faster, the shorter molecules, slower. One particular type of lesion in chromosomal DNA causes it to sediment slower in native sucrose gradients, suggesting that it decreases the molecular weight of duplex DNA. This fragmentation of the chromosome may be due to coinciding interruptions in both DNA strands resulting in "double-strand breaks" (DSBs) (Fig. 5.1A). Although double-strand breaks can be repaired simply by sealing together two DNA ends (the major DSB repair pathway in somatic cells of higher eukaryotes (7.2)), in *E. coli* double-strand breaks are mended by recombinational repair. This conclusion is based on two observations: (1) DSB repair in *E. coli* requires duplicated genomes[1] and (2) DSB repair is not observed in *recA* mutants.[1-3] Chromosome fragmentation may also be due to ruptures within single-strand regions at replication forks (Fig. 5.1B). Restoration of such disintegrated replication forks in *E. coli* is also a function of recombinational repair.

5.1.1. The Origin of Double-Strand Breaks

A common cause of DSBs is ionizing radiation. Chromosomal DNA of *E. coli* analyzed immediately after X-ray or gamma-ray irradiation shows a decrease in its molecular weight proportional to the dose of irradiation.[1,3,4] The mechanism of double-strand break generation by ionizing radiation was discussed in the Introduction (1.2.1). Double-strand breaks in DNA are also formed as a result of replication of a template containing single-strand interruptions. Replication of a DNA molecule with a nick could generate one intact molecule and one broken molecule (Fig. 5.1C). Since any segment of bacterial chromosome, except the terminus, is replicated in one particular direction only (1.3), this way of double-strand break formation in *E. coli* can be realized only in the terminus region.

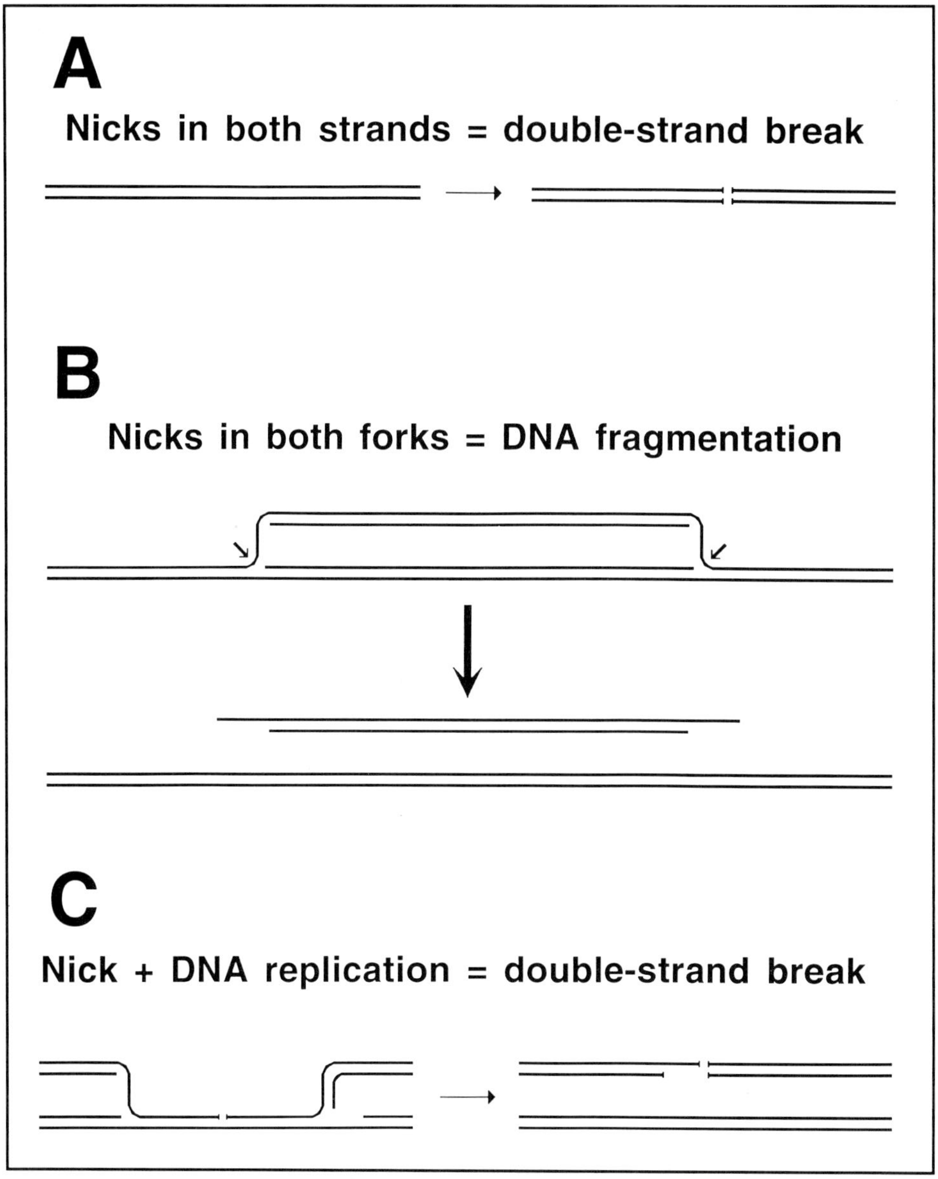

Fig. 5.1. The origin of double-strand ends. (A) Coinciding single-strand interruptions in both DNA strands of a duplex result in a double-strand break. (B) Ruptures in the same DNA strand at both replication forks of a replication bubble lead to chromosome fragmentation. (C) Replication of a template containing a single-strand interruption generates one intact daughter DNA and another daughter DNA with a double-strand break.

DNA transposition can be a source of rare double-strand breaks. Certain transposable elements of *E. coli* (Tn*5*, Tn*7* and Tn*10* among them) excise their genome from donor DNA during transposition to insert it into a new location. Such "conservative" transposition leaves behind a double-strand gap in the donor DNA molecule. The gap is thought to be repaired off the sister chromosome (which still has the element), so the element is preserved in the old location while establishing itself in a new location.[5,6] Still another source of occasional double-strand breaks are endogenous restriction-modification systems. These systems function to modify specific sites in resident DNA and to cut unmodified sites in the incoming, for example, bacteriophage DNA. Sometimes due to modification failure the resident DNA is also cut revealing this as a rare site-specific recombination event.[7] Restriction-modification systems often reside on plasmids, which can be lost from cells harboring them. In this case, slow degradation and dilution of remaining restricting and modifying enzymes ensures that the chromosome will suffer several double-strand breaks, which are repaired in WT cells.[8]

5.1.2. Repair of a Double-Strand Break as Repair of Two Double-Strand Ends. The Origin of Double-Strand Ends

A double-strand break generates a pair of double-strand ends. Since the two ends are likely to have a similar structure, they could be repaired in two equivalent sets of reactions, as proposed in one double-strand break repair model.[9] In this regard, double-strand break repair can be viewed as a sum of two reactions of double-strand end repair, which do not have to be concerted. The alternative possibility is that one end initiates repair and the other one gets involved in the reaction later, when the first end creates a favorable opportunity for the second end.[10] Regardless of whether both ends actually act independently or one end leads while another follows, the central element of any double-strand break repair pathway is likely to be the fate of an individual end.

Besides, there are lesions, such as disintegrated replication forks (see 1.2.1), which comprise a solitary double-strand end. A replication fork is thought to disintegrate when it runs into a preexisting single-strand interruption in template DNA; such interruptions are, for instance, transiently formed during one-strand repair. Also, there are certain *E. coli* mutants (defective, for example, in DNA ligase and DNA polymerase I) which accumulate single-strand interruptions in their DNA. Collapse of replication forks in these mutants may be a regular event, making them dependent on recombinational repair (reviewed in ref. 11).

Inhibition or a complete block of the replication fork progress may also lead to replication fork disintegration. A block could be in the form of a protein bound to a specific site on DNA, or could be caused by malfunctioning of a replisome, or by the absence of DNA precursors (reviewed in ref. 12). Inhibited replication forks in *E. coli* are unstable. There are strains or conditions of growth in which disintegration of inhibited replication forks becomes a routine event, again making cells dependent on recombinational repair.[13,14]

In order to reestablish a disintegrated fork, the solitary double-strand end has to be reattached to the full-length molecule. There are likely to be extensive mechanistic similarities between recombinational repair of double-strand breaks and double-strand ends. This simple-minded assumption allows us to concentrate our discussion on repair of a solitary double-strand end. However, there could be a difference in the structure of double-strand ends created by a double-strand break versus ones resulting from replication fork disintegration.

Double-strand ends generated by a double-strand break are blunt or almost blunt (otherwise, the break would not form). In contrast, a solitary double-strand end created as a result of disintegration of a replication fork might have a long single-strand overhang of either polarity. Indeed, if a single-strand interruption is in template strand for the leading strand DNA synthesis, the last primer is unlikely to be used up because of the disruption of a replisome, leaving the 5'-ending tail down from this primer single-stranded. If a single-strand interruption is in template strand for the lagging strand DNA synthesis, the 3'-overhang cannot be filled in to the end, because DNA synthesis has the 5'-to-3' directionality. Thus, a solitary double-strand end released after replication fork disintegration is likely to have a single-strand tail of either polarity with an average length about half the length of an Okazaki fragment (1.3.1), that is, 750-1000 nucleotides.

5.2. AN OUTLINE OF DOUBLE-STRAND END REPAIR

Although double-strand end repair has been known about for a long time, only the basic principles of this recombinational repair reaction are understood. There are still genes whose inactivation severely compromises mending of double-strand breaks in *E. coli* and whose products are yet to be characterized biochemically. However, certain fragments of this complex picture are detailed quite well and will be discussed later in the chapter. But first, a general outline of double-strand end repair is necessary, since new enzymatic activities, important in recombinational repair, need to be introduced.

5.2.1. Presynaptic Phase of DSE Repair: Helicases and Nucleases Make a Blunt End Single-Stranded

A double-strand break is a result of the coincident scissions in the two DNA strands, generating two blunt or nearly blunt double-strand ends. Sometimes these two ends can be sealed back together by DNA ligase. The "simple ligation" option is often used by eukaryotes (reviewed in ref. 15), where the complex chromosomal packaging is unlikely to allow two ends of the break to move far away from each other.[16] However, the two ends of a double-strand break in a prokaryotic chromosome are relatively loose, so the simple ligation option is unfeasible. In order to bring the two loose ends together, an intact homologous sequence is used.

Homologous pairing of a broken sequence and an intact sequences could be accomplished by RecA protein. Since RecA protein does not polymerize on a duplex DNA in physiological conditions (2.2.1), blunt ends need to be made single-stranded to permit RecA polymerization. Blunt ends can be made single-stranded in two principally different ways, which are often combined in vivo (Fig. 5.2). One way is to unwind two DNA strands from the end and complex them with SSB protein to prevent their reannealing. The unwinding of DNA is a function of enzymes called DNA helicases. There are twelve characterized DNA helicases in *E. coli*, some of them participating in DNA replication, others being employed in DNA repair. Their apparently similar in vitro DNA unwinding activities could be targeted in vivo to diverse DNA structures by SSB protein.[17]

The other principal way to make a duplex DNA end single-stranded is to eliminate one of the DNA strands. This is a function of another abundant class of enzymes — DNA exonucleases, which degrade DNA

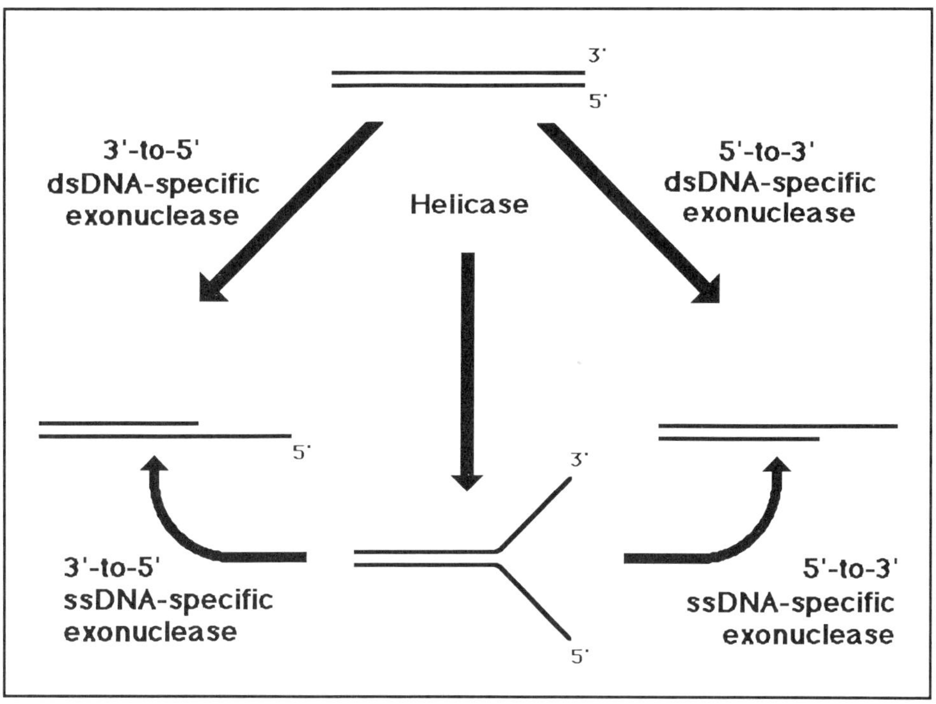

Fig. 5.2. Processing of a blunt double-strand end by helicases and nucleases. Explanations in the figure and in the text.

molecules having free ends. Some exonucleases degrade both DNA strands simultaneously which is hardly helpful in repair. Other exonucleases are specific for a particular DNA strand. Exonucleases degrading the DNA strand ending 3' at the end generate 5'-overhangs, while exonucleases degrading the DNA strand ending 5' at the end generate 3'-overhangs (Fig. 5.2). Linear DNA with single-strand overhangs is also produced if one of the single strands unwound by a DNA helicase is degraded by a single-strand DNA-specific exonuclease of a particular polarity (Fig. 5.2). At last, as discussed above, double-strand ends produced as a result of replication fork disintegration, may already have single-strand overhangs of a substantial length. But even in this case they might need an additional exonucleolytic processing (see 5.6.1).

5.2.2. RecA Assembles on Double-Strand Ends in Two Opposite Directions Depending on the Polarity of Single-Strand Tails

Removal of either 3'-ending or 5'-ending strand starting from a blunt end would allow RecA to assemble its filament on a resulting ss tail. Since RecA filament assembles and treadmills along ssDNA in the 5' to 3' direction,[18] it would migrate into the duplex portion of the end with a 5'-overhang, while concentrating at the tip of the end with a 3'-overhang without venturing into the adjacent duplex region (Fig. 5.3). The polymerization of RecA filament on the single-strand overhang of a double-strand end signals the onset of a homologous search for the intact sister sequence. When such a sequence is found, RecA filament is believed to align it with the double-strand end and to begin strand invasion (Fig. 5.3). For the strand invasion to happen, two topological reactions should be catalyzed (4.4). First, a particular degree of unwinding (negative superhelicity) in the intact homologous sequence has to be maintained by DNA gyrase. Second, the incoming single-strand DNA has to be wound around the resident complementary strand by DNA topoisomerase I.

The two possible ways of RecA filament assembly generate two distinct patterns of double-strand end invasion, in which either the 5'-ending strand or the 3'-ending strand invade first. The opposite patterns of double-strand end invasion result in two distinct configurations of Holliday junctions in which the strands invading first are crossed while the strands invading second are continuous (Fig. 5.4). Thus, it seems possible to determine which situation is realized in vivo by analysis of heteroduplex polarity in recombinant chromosomes. Such an analysis done on a particular locus in phage lambda shows that 3'-overhangs are more frequent than 5'-overhangs in the recombination products associated with crossing-over,[19] while in the noncrossover recombinants the donated information is usually found in the opposite strand.[20] The schematic diagram of how both such products could be formed in a single resolution event is presented in Fig. 5.4.

In this diagram, the Holliday junction is drawn in its open planar configuration as well as in both cross-over configurations, with the preferential way of resolution indicated by small arrows. The picture is to illustrate that such an analysis provides meaningful information about the polarity of invasion only if we assume that the Holliday

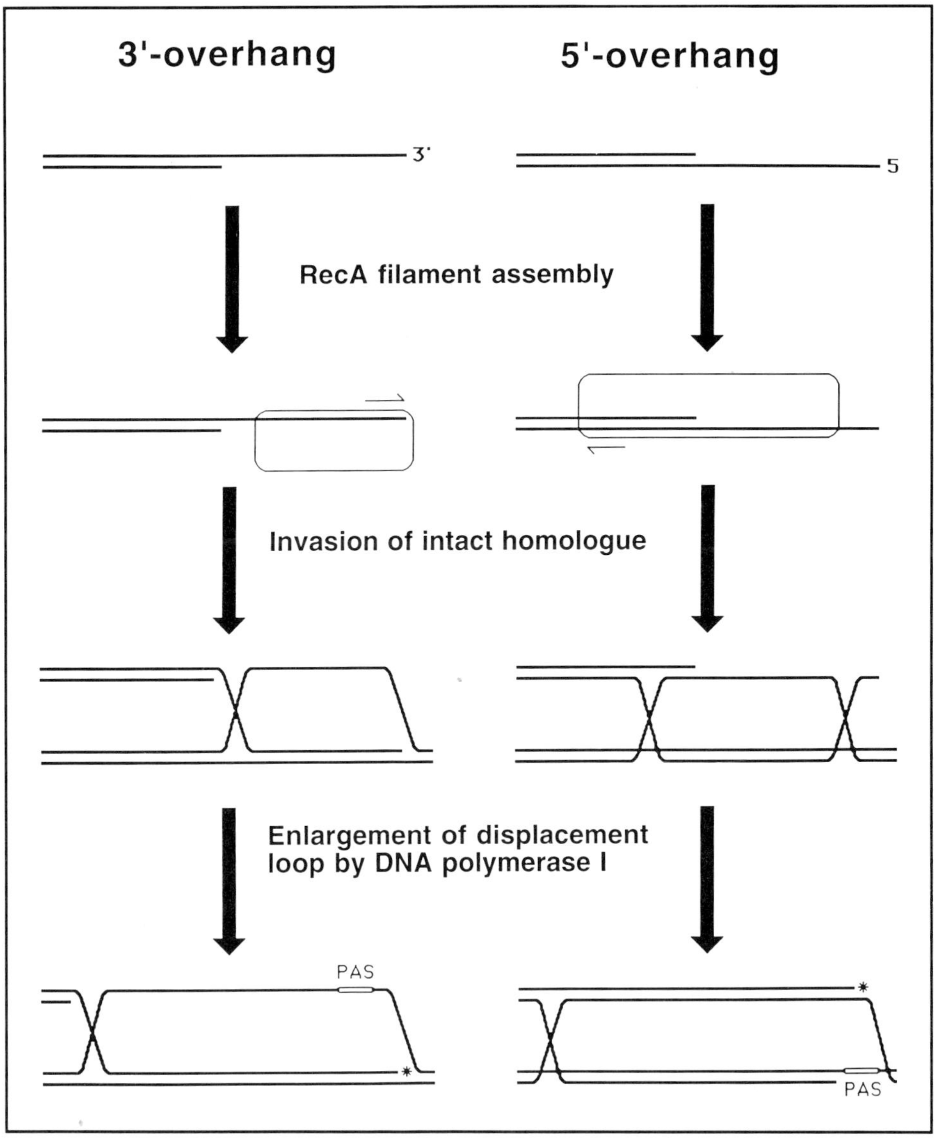

Fig. 5.3. Overhangs of the opposite polarities lead to distinct patterns of RecA-catalyzed double-strand end invasion. RecA filament is shown as open rectangle with rounded corners. Small one-sided arrow indicates the direction of RecA filament assembly. PAS – primosome assembly site; star – limited DNA synthesis by DNA polymerase I. Explanations in the figure and in the text.

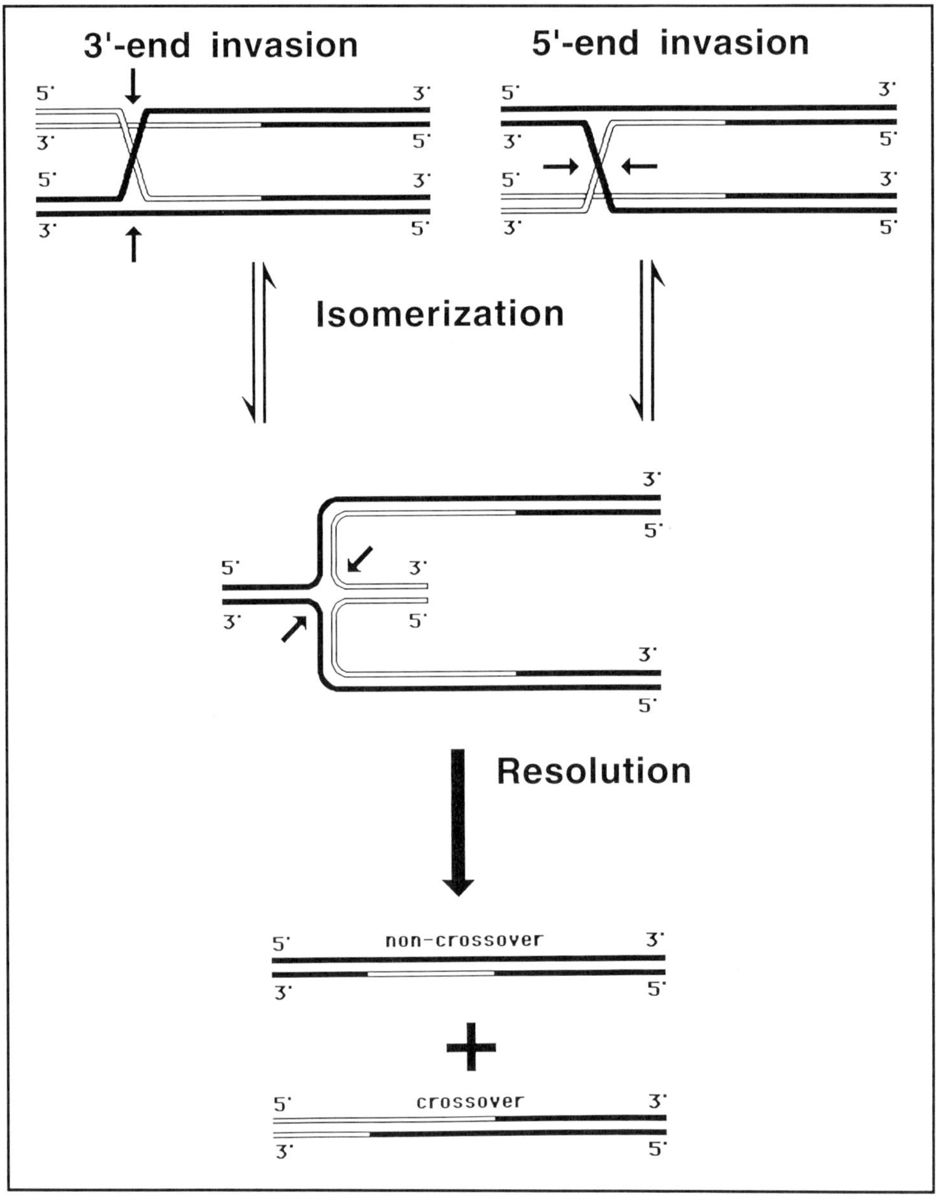

Fig. 5.4. Preferential resolution of Holliday junctions as revealed by heteroduplex analysis of recombinant chromosomes. The invading (white) strands are extended by DNA synthesis in order for the final products be complete. It is clear that both polarities of invasion (top panel) are able to account for the observed recombinant chromosomes (bottom panel). Moreover, the two configurations of Holliday junctions can isomerize into each other, further complicating the analysis.

junction generated during double-strand end invasion: (1) is not moved around extensively; (2) does not isomerize and (3) is resolved in a fixed direction. Since none of these three assumptions is confirmed in vivo, for the time being both pathways, initiated by either 3'- or 5'-overhangs, should be treated as equally probable (Fig. 5.3). However, we know that RecA polymerization on SSB-complexed single-strand DNA requires assistance of other proteins, which are most probably targeted to the junction of single-stranded and duplex DNA. If we assume these conditions, the 5'-to-3' polarity of RecA filament assembly dictates that only ends with 3'-overhangs will be invasive. This preference will be reinforced later, during discussion of specific mechanisms of the presynaptic phase of double-strand end repair.

5.2.3. Postsynaptic Phase of DSE Repair: Regeneration of a Replication Fork and Processing the Holliday Junction

RecA-catalyzed synapsis of a double-strand end with an intact homologous duplex opens a possibility for regeneration of a replication fork. Initiation of DNA synthesis in *E. coli* requires structure-specific binding of a primosome-loading protein, either DnaA or PriA, near or within the single-strand region (1.3). Since DnaA or PriA binding sites (called PAS, or primosome-assembly sites) are probably rare, there is likely to be an extension of the displacement loop before such a site in the displaced single-strand DNA is exposed (Fig. 5.3). This enlargement of the displacement loop could be a result of a limited DNA synthesis, primed at the invading 3' end and carried out by DNA polymerase I (Fig. 5.5) (5.4.1). In this scheme, the initiation of DNA replication after invasion of a double-strand end is similar to initiation of plasmid DNA replication[21] (1.3.3).

RecA-mediated strand invasion by a 5'-ending overhang generates a regular Holliday junction (Fig. 5.3). RecA-promoted strand invasion by a 3'-overhang results in a three-strand junction, which should be free from RecA (Fig. 5.3). Although this three-strand junction may be accessible to cleavage by RuvC resolvase, which in vitro is able to cleave such junctions,[22] this "early" resolution is unlikely to be of importance due to sequence specificity of RuvC resolution.[23] Most probably, DNA synthesis initiated on the displaced strand converts the three-strand junction into a Holliday junction (Fig. 5.5C and D). Resolution of this Holliday junction behind the re-created replication fork and removal of the used RecA filament, catalyzed for example by RuvAB + RuvC enzymes (4.5), would complete the restoration of a disintegrated replication fork (Fig. 5.5E) (5.4).

5.3. RecBCD Nuclease Prepares Double-Strand Ends for RecA Polymerization

In WT *E. coli*, rejoining of the chromosomal DNA fragmented by gamma-irradiation is completely blocked by only three single

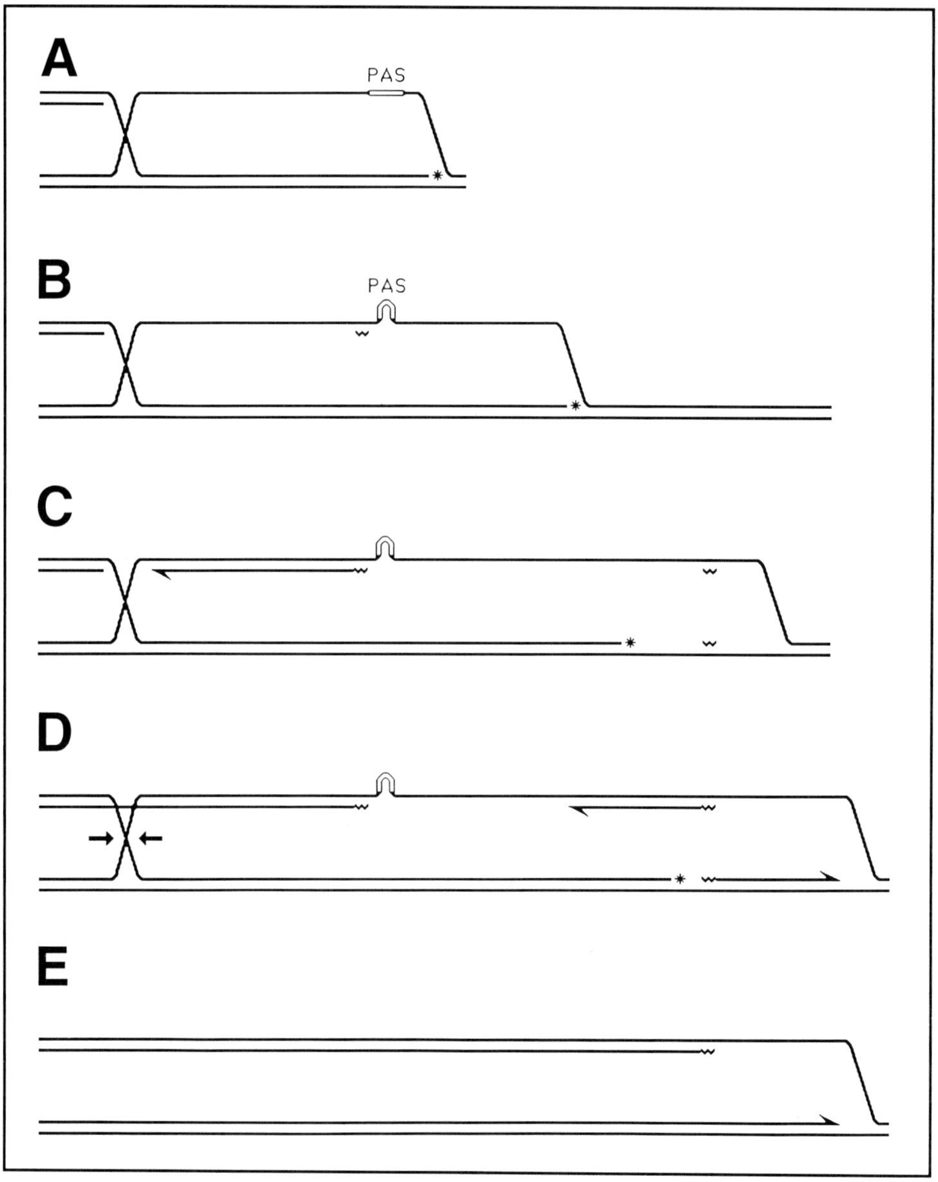

Fig. 5.5. Postsynaptic phase of double-strand end repair. PAS – primosome assembly site; star – limited DNA synthesis by DNA polymerase I; little wavy segments – RNA primers. (A) RecA-catalyzed invasion of the 3'-end allows limited DNA synthesis to be primed by DNA polymerase I. As a result of this synthesis, a primosome assembly site is unraveled in the displaced strand. (B) PriA assembles a primosome at PAS, which lay down primers and unwinds DNA at the fork (C). (D) A replisome is loaded onto the primed replication fork; the three-strand junction is converted into a Holliday junction. (E) Double-strand end repair is completed with the resolution of the Holliday junction (the direction of the resolution is shown in (D)).

mutations:* *recA*, *recB*, and *recC*.[26] Above (5.2.1) it was discussed how a double-strand end needs to be acted upon by a helicase, a nuclease, or a combination of both to tailor it for RecA polymerization. RecB and RecC proteins associate with RecD protein to form RecBCD enzyme, which is a combination of a potent helicase with duplex DNA- and single-stranded DNA-specific exonucleases. In vitro, RecBCD enzyme degrades linear duplex DNA to oligonucleotides with an incredible speed. In vivo, it degrades bacteriophage DNA cut by the host restriction systems. Also, in *recA* mutants, RecBCD nuclease degrades the entire chromosome after it has been fragmented as a result of DNA damage. Efficient degradation of DNA by RecBCD nuclease is difficult to reconcile with the central role this enzyme plays in double-strand break repair. *E. coli* employs a two-component system to turn this "Mr. Hyde" of bacterial cells into "Dr. Jekyll" when there is a need to repair double-strand ends.

5.3.1. RecBCD Enzyme: Genes and Mutants

RecBCD nuclease is encoded by three genes, *recB*, *recC* and *recD*, closely positioned on the *E. coli* chromosome. *recB* and *recD* genes form an operon, while *recC*, although situated nearby, has its own promoter.[27-31] The combination of the extremely weak promoters with the use of suboptimal codons ensures no more than 10 RecBCD enzyme molecules per chromosome.[32] None of its three genes is SOS-inducible.

There are two major phenotypes displayed by mutants in *recB*, *recC*, and *recD* genes. Null mutants in *recB* or *recC* genes are sensitive to DNA damaging treatments[33] and have a viability around 30%.[34] Although the repair of daughter-strand gaps in these mutants is normal, they die after their DNA has been damaged, apparently because of their inability to repair double-strand breaks.[35] Other possible reasons for their low viability will be discussed later (5.5). In contrast, null *recD* mutants and a single *recC* missense mutant display normal viability as well as the wild-type survival after DNA damaging treatments.[28,36] Null mutants of both groups lack powerful exonuclease activity of RecBCD enzyme, although *recD* cells still retain half the DNA degradation capability of WT cells.[37]

5.3.2. RecBCD Enzyme: Biochemical Activities

RecBCD enzyme is a heterotrimer[38] of RecB (134 kDa), RecC (129 kDa) and RecD (67 kDa) polypeptides (for review see ref. 32). The prominent activities of RecBCD enzyme include helicase, dsDNA

* ssb *mutants are not tested yet, although they are also likely to be deficient in* recA-*dependent rejoining of fragmented chromosome, since SSB protein is required for efficient RecA filament assembly (2.2.5).* ssb *mutants are defective in recombination requiring double-strand end repair.*[24,25]

exonuclease and ssDNA exonuclease. The hallmark of RecBCD nuclease is that it requires ATP hydrolysis for its nucleolytic activities. From a chemical point of view, the hydrolysis of DNA does not need input of energy, the point best illustrated by the fact that most of the nucleases hydrolyze DNA in the absence of high-energy cofactors like ATP. In fact, RecBCD nuclease of *E. coli*, also called ExoV,[39] and analogous enzymes of other prokaryotes are the only known ATP-dependent exonucleases.[32,40,41] No enzymes analogous to prokaryotic ExoV has been found in eukaryotic cells. The wealth of biochemical information about RecBCD nuclease of *E. coli* reveals the mechanism of action of this unique enzyme.

ssDNA exonuclease activity is defined as the one able to degrade linear single-strand DNA (ssDNA). RecBCD degrades ssDNA down to pieces several nucleotides in length.[39,42,43] This reaction needs ATP, but is insensitive to ATP concentration; hence, ATP is thought to be used as an allosteric effector.[44] dsDNA exonuclease activity is defined as degrading linear double-strand DNA. RecBCD degrades linear duplex DNA down to the same oligonucleotide products as ssDNA, but with a greater speed.[39,42,43] Circular duplex DNA, even containing single-strand interruptions and short gaps, is refractory to RecBCD attacks[39,43] because the enzyme can enter duplex DNA only from double-strand ends.[45]

The degradation of duplex DNA by RecBCD absolutely requires both Mg^{2+} and ATP, the rate of degradation being dependent on the ratio of these two ions. ATP and Mg^{2+} complex each other, so in their equimolar solution there is little free magnesium and little free ATP (ref. 46 and references therein). The nuclease activity of RecBCD is stimulated when the concentration of Mg^{2+} ions exceeds that of ATP, and is inhibited when ATP concentrations are higher than that of Mg^{2+}.[47,48]

The third major activity of the RecBCD enzyme is DNA unwinding. RecBCD is a potent and highly processive DNA helicase, which in vitro can unwind DNA at a rate close to 1000 bp/second,[49] unwinding on average 30 kilobase pairs per binding event.[50] The DNA unwinding by RecBCD absolutely requires ATP, which is not surprising, because dissociation of paired DNA strands does require input of energy. In fact, all the helicases discovered to date require ATP to catalyze DNA unwinding. The logical explanation for the unusual ATP-dependent dsDNA exonuclease activity of RecBCD is that the enzyme is able to hydrolyze duplex DNA only after unwinding it (discussed in ref. 41).

5.3.3. RecBCD Enzyme: Mechanism of DNA Hydrolysis

The aminoacid sequences of both RecB and RecD contain ATP-binding domains,[29,30] and both proteins bind ATP.[51] Moreover, RecB subunit can hydrolyze ATP[52] and displays a weak helicase activity, translocating in the 3' to 5' direction along ssDNA.[53] Although RecD itself is not an ATPase, inactivating its ATP-binding domain results in inactivation of the duplex DNA exonuclease activity of the reconstituted RecBCD enzyme.[52]

In vitro, RecBCD degrades DNA asymmetrically — the 3'-ending strand receives most of the cuts, while the 5'-ending strand receives a few.[54] This asymmetry of degradation reflects the asymmetry of the enzyme binding to dsDNA. In the absence of ATP but in the presence of Mg^{2+}, RecBCD binds to a double-strand end, but does not proceed into the DNA molecule. The pattern of DNA protection by such a bound enzyme suggests that RecB subunit binds to the 3'-ending strand, while RecC and recD subunits sit on the 5'-ending strand[55] (Fig. 5.6A). In the presence of ATP, Mg^{2+}, and Ca^{2+} (to inhibit the nuclease activities of the enzyme), RecBCD translocates along the DNA, unwinding it and forming a growing loop on the 3'-ending strand. In vitro, the size of the loop is approximately 1/3 of the length of the unwound region.[56] The loop is postulated to exist in vivo and may be an important regulatory element for the RecBCD enzyme. In conclusion, a DNA end being processed by RecBCD probably has a single-strand loop on the 3'-ending strand held by the enzyme, with little or no DNA past the enzyme, while its 5'-ending strand hangs as an unpaired tail and is periodically clipped by the enzyme[32,41] (Fig. 5.6B).

5.3.4. Chi-Sites: Signal Sequences for the RecBCD-Promoted Recombinational Repair

RecBCD enzyme interacts with an asymmetric short DNA sequence called Chi. Chi-sites were discovered in phage lambda as mutations creating hot-spots for recombination (for review see ref. 57). If phage lambda fails to inactivate ExoV activity of RecBCD enzyme with its Gam protein and lacks its own recombinational genes, its DNA is savaged by RecBCD when linearized for packaging by a phage-encoded enzyme.[58] However, when a point mutation creates a Chi-site in such lambda, phage DNA linearization does not lead to its complete degradation but results in recombination with other lambda DNA molecules catalyzed, surprisingly, by the same RecBCD enzyme. A solution for the paradox of a single enzyme carrying out two mutually-exclusive activities on linear DNA (complete destruction vs. preservation through recombination) was suggested on the basis of genetic data. It was proposed that when RecBCD enzyme encounters a Chi-site, its dsDNA exonuclease activity is disabled. Since the modified enzyme still retains its helicase activity, it becomes a "recombinase".[59] This idea has received support from in vitro experiments, as discussed below. It was also shown that Chi-sites prevent the rapid degradation of linear DNA in vivo,[60,61] further substantiating the hypothesis that they act as signal sequences, converting RecBCD from nuclease to recombinase.

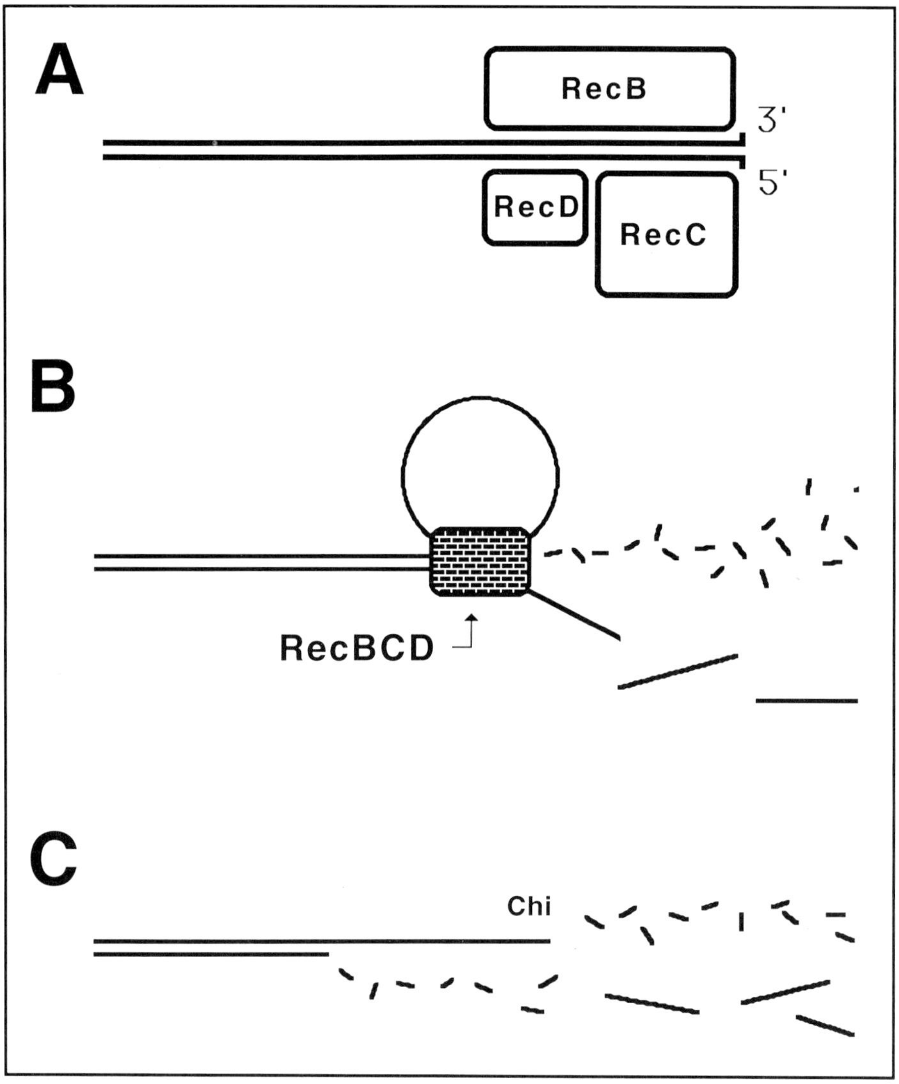

Fig. 5.6. The mode of RecBCD enzyme binding to a DNA end and the patterns of RecBCD-promoted DNA hydrolysis. (A) The scheme of RecBCD binding to a double-strand end. (B) The mode of RecBCD action on Chi-less DNA. (C) The mode of RecBCD action in the vicinity of a Chi-site. The direction of enzyme travel in B and C is from right to left.

5.3.5. Chi-Sites: Structure and the Mechanism of Action

In *E. coli*, Chi-site is an asymmetric octanucleotide sequence 5'-GCTGGTGG-3';[62] Chi-sites of other bacteria are likely to be different.[63] Consistent with its asymmetry, recognition of Chi by RecBCD occurs only when the enzyme approaches the site from the 3'-side, but not from the 5'-side.[62,64-66] The probability of Chi-site recognition by a RecBCD molecule is less than 50% both in vivo and in vitro.[59,61,67,68] When RecBCD sees Chi, it makes a nick several nucleotides to the 3'-side of the Chi sequence.[66,69] It was assumed for some time that the nicking is relevant to the role of Chi-sites in recombination, since the level of nicking of Chi-like sequences (different in a single position from the canonical octanucleotide) correlates well with the ability of the Chi-like sequences to stimulate recombination.[70] However, eventually it was found in vitro that RecBCD travels along the DNA cutting it at some regular intervals — the higher the speed of the enzyme progress, the longer the pieces of DNA produced. It was also found that the enzyme pauses at a properly oriented Chi-site. After some time, RecBCD resumes its movement along the DNA, but now it cuts DNA infrequently, at least the 3'-ending strand.[48,54] Thus, an alternative interpretation of nicking at Chi is that: (1) while the enzyme has a pause at Chi, it delivers its regular cut and (2) this is the last cut of RecBCD, since after passing Chi, it loses its ability to cut.[48,54] According to this interpretation, nicking at Chi per se is not needed for the increased recombination promoted by Chi-sites.

This interpretation is in accord with characteristics of general recombination in *recD* null mutants. The RecBC(D$^-$) mutant enzyme acts as if it is permanently modified by Chi-sites,[71] suggesting that at Chi-site, RecBCD loses its RecD subunit (or the RecD subunit gets modified), which turns a destructive enzyme into a recombinational repair enzyme.[59,71,72] This idea is consistent with in vivo findings that overproduction of RecD polypeptide decreases Chi effect as hot-spot for general recombination[73-75] and compromises cell survival after ionizing radiation.[76] In vitro experiments indicate that the RecD subunit of the enzyme may indeed regulate the nuclease function of RecBCD when the enzyme passes a Chi-site.[48,77] Recently, it was realized that in vitro, RecBCD enzyme may actually change the polarity of strand degradation upon passing a Chi-sites:[48] before the Chi-site, it degrades mostly the 3'-ending strand, while after the Chi-site, it degrades mostly the 5'-ending strand (Fig. 5.6C). Thus, after passing a Chi-site, the enzyme seems to generate a 3'-single-strand overhang.

5.3.6. Chi-Sites: RecA Connection?

Inhibition of the nuclease activities of RecBCD enzyme by Chi in vitro does not require any other protein.[54,68,77] By contrast, disabling of the nuclease activity of RecBCD in vivo requires RecA protein.[60,61] Besides all the trivial explanations for this RecA requirement, there is an intriguing possibility that RecA is actually needed for the modification of RecBCD at Chi in vivo.

This possibility is indirectly supported by the fact that the optimal recombinational repair in vivo is achieved only when RecA and RecBCD enzymes are from the same or closely related species. Both

RecA and RecBCD enzymes are ubiquitous in bacteria, allowing complementation studies in *E. coli* mutants, deleted for their own *recA* and *recBCD* genes. If such *E. coli* mutants are provided with RecA and RecBCD from the *E. coli* relative, *Serratia marcescens*, or from a more distantly related microbe, *Proteus mirabilis*, their recombinational repair proficiency is restored to 90% and to 60% of the WT *E. coli*, respectively.[78] When RecA protein of *E. coli* is asked to work with the RecBCD enzyme from *S. marcescens*, or vice versa, the resulting strains still boast almost WT recombinational repair proficiency. However, if RecA protein is provided from either *S. marcescens* or *E. coli*, while RecBCD enzyme is from *P. mirabilis*, or vice versa, the resulting strains show only 20-50% recombinational capabilities compared to WT *E. coli*. The poor interspecies complementation is interpreted to signify cooperation between RecA and RecBCD enzymes.[78]

Direct interactions between RecA and RecBCD would also provide a clue to the observed strand preference of RecA polymerization on DNA, acted upon by RecBCD. RecA polymerization on single-strand DNA is inhibited in the presence of SSB (3.2). In the daughter-strand gap repair, RecA polymerization on SSB-bound single-strand DNA is facilitated by RecF, RecO and RecR proteins (4.2). No protein facilitating RecA polymerization has been described for the RecBCD pathway yet. The most economical suggestion is that RecBCD itself stimulates RecA polymerization on SSB-covered single-strand DNA.

Which DNA strand is chosen by RecA if its assembly is promoted by RecBCD? This does not seem to be an important question anymore since after passing a Chi-site, RecBCD apparently leaves a 3'-single-strand overhang.[48] However, in vitro results indicate that even in conditions preventing RecBCD degradation of the 5'-ending strand, the 3'-ending strand is favored by RecA for the joint molecule formation.[48,79] Since under these reaction conditions RecBCD completely unwinds duplex DNA, so that RecA could polymerize on either strand, the preferential polymerization of RecA on the 3'-ending strand suggests that it is actively promoted by RecBCD.

In vitro, RecBCD seems to loose its RecD subunit at Chi-sites without assistance from other proteins, while in vivo, the same modification of RecBCD might require RecA protein (see above). It is tempting to speculate that not only does RecBCD help RecA to polymerize, but in return, RecA polymerization triggers RecBCD removal from DNA after the enzyme has encountered a Chi site. In fact, RecBCD removal from DNA after Chi in the absence of RecA could be disadvantageous for the cell, because in situations when recombinational repair is overloaded and RecA is in short supply, the DNA degradation offers a life-saving alternative (5.5). Thus, the cells might control RecBCD modification at two levels: (1) by Chi sites and (2) by RecA polymerization.

5.4. TWO PATHWAYS FOR THE POSTSYNAPTIC PHASE IN THE DOUBLE-STRAND END REPAIR

Above it was stated that rejoining of double-strand breaks in WT *E. coli* cells measured as reassembly of fragmented chromosome after treatment with ionizing radiation requires only RecA and RecBCD enzymes. This is analogous to the situation with daughter-strand gap repair, where after UV-irradiation, sealing the daughter-strand gaps depends only on RecA, RecF, RecO and RecR proteins (4.2). As we know, the viability of *E. coli* cells after exposure to UV also depends on *ruvA*, *ruvB* and *ruvC* proteins (4.5). Similarly, the viability of *E. coli* cells after exposure to ionizing radiation is compromised by several additional mutations: *polA*, *priA*, *ruvA*, *ruvB*, *ruvC*, *recG*, *recN*, and *rorB*, although some of these mutants (*polA*, *priA* and *recG* were not tested) show almost normal ability to rejoin fragmented chromosome after low doses of ionizing radiation.

5.4.1. DNA POLYMERASE I AND PRIA: FORMATION OF A REPLICATION FORK

RecBCD- and recA-catalyzed double-strand end invasion into an intact duplex generates a replication fork framework. First, a primosome, and then a replisome must be loaded onto this framework to convert it into a functional replication fork. A primosome is loaded at a primosome assembly site if it is exposed in the displaced single-stranded loop. To unravel a primosome assembly site, a limited DNA synthesis primed by the 3'-invading end might be needed (Figs. 5.3 and 5.5A). This DNA synthesis is likely to be catalyzed by DNA polymerase I, a small one-subunit repair polymerase of *E. coli* coded by *polA* gene.[80,81] DNA polymerase I has three activities: 5'→3' polymerization activity, 3'→5' exonuclease (proofreading) and unusual 5'→3' exonuclease, which allows the enzyme to replace segments of damaged DNA or RNA with regular DNA "in one stroke" (reviewed in ref. 82). *polA* mutants are sensitive to both UV and ionizing radiation[81,83] (primarily because of their defect in one-strand repair). Polymerization-deficient *polA* mutants also show a 10- to 20-fold decrease in recombination requiring double-strand end repair, but a *polA* mutant deficient in 5'→3' exonuclease is not defective in such recombination.[25,84] This is consistent with our scheme, according to which only the polymerization function of DNA pol I is important in double-strand end repair.

Following the limited displacement synthesis by DNA pol I, the replication fork restart is believed to begin at a primosome assembly site (PAS) exposed on the displaced strand, converting this strand into a template for the lagging strand DNA synthesis (Fig. 5.5A).[85] The primosomal n' protein, also called PriA, binds to PAS and promotes primosome loading onto it (1.3.3). *priA* mutants are sensitive to DNA damaging agents and deficient in homologous recombination requiring double-strand end repair.[86,87] Certain point mutations in *dnaC*, a factor which delivers the DnaB replicative helicase to the replication fork, suppress both defects of *priA* mutants, suggesting that they are caused by the inability of *priA* mutants to reload the primosome.[86,87] The two principal components of a primosome are DnaB helicase, which drives replication fork unwinding, and DnaG primase, which lays the primers for DNA synthesis (1.3.1) (Fig. 5.5B). The availability of primers signals that a replisome can be loaded on the replication fork framework, finishing the regeneration of a replication fork. Now, to complete double-strand end repair, the DNA junction and the used RecA filament are to be removed. In the daughter-strand gap repair, RuvA, RuvB and RuvC enzymes clean up after the completion of repair (4.5). In the double-strand end repair, besides RuvABC team there is a competing team doing the same job.

5.4.2. The Roles of RuvAB, RecG and RuvC in the RecBCD Pathway

The two tasks of RuvA and RuvB proteins during daughter-strand gap repair are believed to be the enhancement of the RuvC resolution activity and displacement of spent RecA filaments from DNA (4.5). These two activities, which could be related, stem from the ability of RuvAB to bind and move Holliday junctions, even still within RecA filaments. As a result of a daughter-strand gap repair reaction, two such junctions are proposed to be formed (4.5). The importance of the RuvAB-catalyzed step in daughter-strand gap repair is deduced from the poor survival of *ruvA* or *ruvB* mutants after UV irradiation.[88,89] After double-strand end invasion, a single junction is formed, which can be either three-stranded or four-stranded (Fig. 5.5). Interestingly, *ruvA* or *ruvB* mutants are only moderately defective in recombination after conjugation or transduction,[88-90] the two recombination reactions that require double-strand end repair. This suggests that another activity can substitute for RuvAB in the removal of DNA junctions and used RecA filaments during double-strand end repair.

Indeed, the mild defect of *ruv* mutants in conjugational recombination is aggravated by a *recG* mutation.[90] A *recG* mutation by itself causes a moderate reduction in cell survival after UV- or X-ray treatment.[90,91] *recG* is the last gene of the *spoT* operon, encoding a 76 kDa protein; there is no indication that a low-level expression of *recG* is enhanced during the SOS response.[92] The synergism of *ruv* and *recG* mutations is nicely accounted for by the fact that *recG* encodes another DNA helicase, whose in vitro activities seem to completely overlap those of RuvAB enzyme. RecG helicase binds Holliday junctions and drives their branch migration, even through regions of nonhomology.[93] It can also dissociate RecA-made joint molecules containing Holliday junctions, even those still covered by RecA filament.[94]

5.4.3. An Alternative Way to Remove a Junction and to Restore a Replication Fork

The in vitro activities of RecG suggest that it may substitute for RuvAB in RuvC-promoted junction resolution. If this is the role of RecG, *ruvC* mutants must be more sensitive to DNA damage than are *ruvAB* mutants. On the contrary, *ruvC* mutants are no more UV-sensitive and recombination-deficient than *ruvAB* mutants, while *ruvC recG* doubles are as deficient as *ruvAB recG* doubles.[90] This indicates that: (1) RecG does not substitute for RuvAB in vivo and (2) there must be an alternative way of resolving DNA junctions, which is independent of RuvAB + RuvC, but dependent on RecG.

If the RecG-dependent pathway is to work by a mechanism similar to that of the RuvAB + RuvC pathway, there must be another RuvC-like resolvase in *E. coli* cooperating with RecG. Indeed, RusA was found to be a resolvase,[95] which dramatically improves the ability of *ruv* mutant cells to repair DNA damage in a RecG-dependent fashion.[96] However, RusA is not normally expressed and needs to be activated by mutation, making its participation in the RecG-dependent pathway in WT cells unlikely. The absence in *E. coli* of canonical resolvases other than RuvC hints to a possibility that the RecG pathway employs a different mechanism of junction removal.

For example, a Holliday junction generated by a double-stranded end invasion can be removed if one of the branching strands at the D-loop is incised, and the D-loop-distal end of the nick is ligated to one of the invading strands (Fig. 5.7). Generation of a double-stranded end near the Holliday junction allows removal of the junction simply by translocating it to the free end. Besides resolution, sliding the junction off this end removes used RecA filament and simultaneously creates a replication fork framework (Fig. 5.7). Thus, if the bubble generated as a result of double-stranded end invasion is opened by incision, no Holliday junction resolution by RuvC is needed.

5.4.4. The RecG Pathway May Handle Junctions Mostly at the Three-Strand Stage

The proposed alternative way of removing junctions works for three-stranded junctions as well. In this case, however, the incision at the D-loop needs to be strand specific — to result in a replication fork restoration, the branching strand must be cut. The position of the nick that "resolves" a three-stranded junction, is now not restricted to the D-loop. An alternative scenario can be drawn in which the 3'-invading end (Fig. 5.8A) primes limited DNA synthesis by DNA polymerase I (Fig. 5.8B), while RecG pushes the three-stranded junction in the same direction, removing the invading strand from the D-loop and dispersing the RecA filament (Fig. 5.8C). After the RecA filament is completely removed, the bare three-stranded junction can be incised at the continuous strand, which would restore the replication fork framework (Fig. 5.8C-E). Recombination aficionados might recall that both modes of resolution of three-stranded junctions (Figs. 5.7 and 5.8) were proposed by Fox in 1966[97] (1.1.4).

The suggested mechanism for RecG-dependent three-stranded junction removal is compatible with one of the few in vitro differences between RecG and RuvAB. While both enzymes are able to translo-

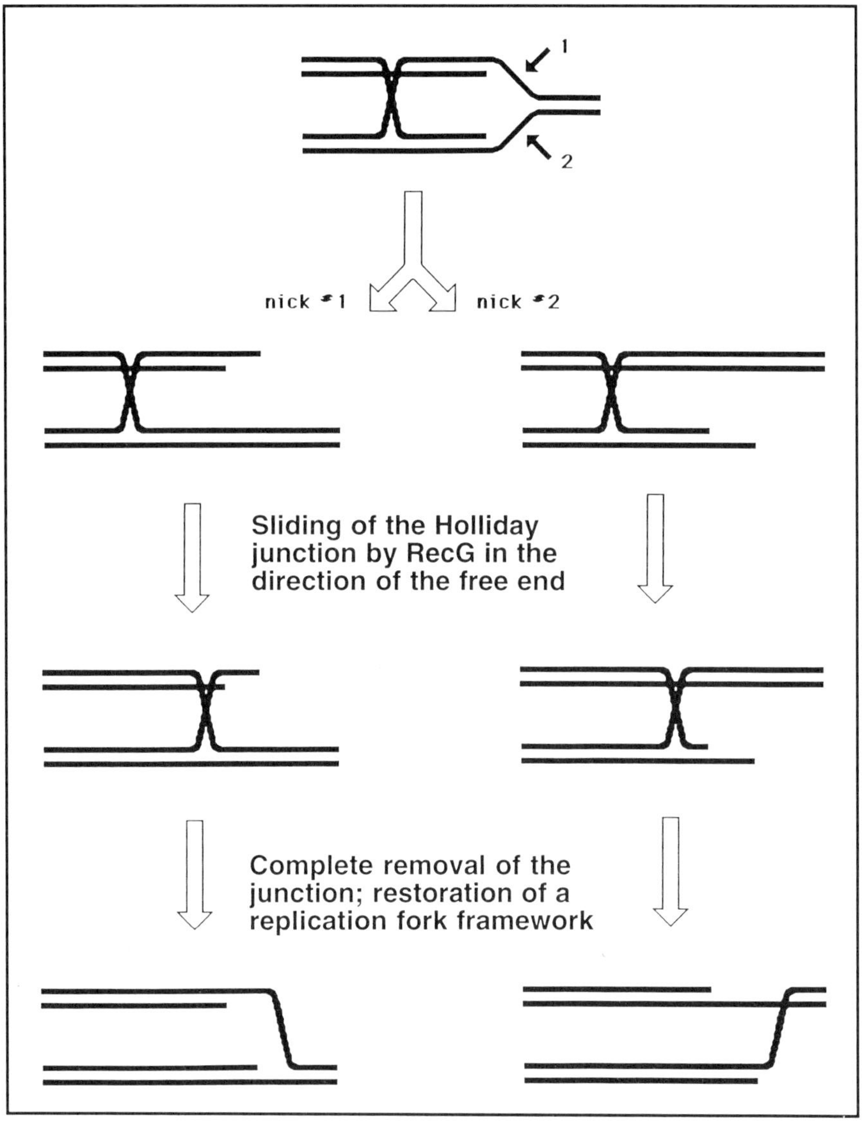

Fig. 5.7. An alternative way to remove a Holliday junction associated with a replication fork framework. Small numbered arrows indicate position of the single-strand scissions. Explanations in the figure and in the text.

cate deproteinized three-strand junctions, only RecG can translocate them when such junctions are still in the grips of RecA filament.[98] In doing so, RecG disrupts RecA-promoted pairing, dissociating the joint molecules. Interestingly, the directionality of this reaction seems to be determined by RecA filament itself — when RecA is absent, RecG prefers to translocate the junction in the opposite direction.[98] These observations suggest that in vivo RecG could act at three-strand

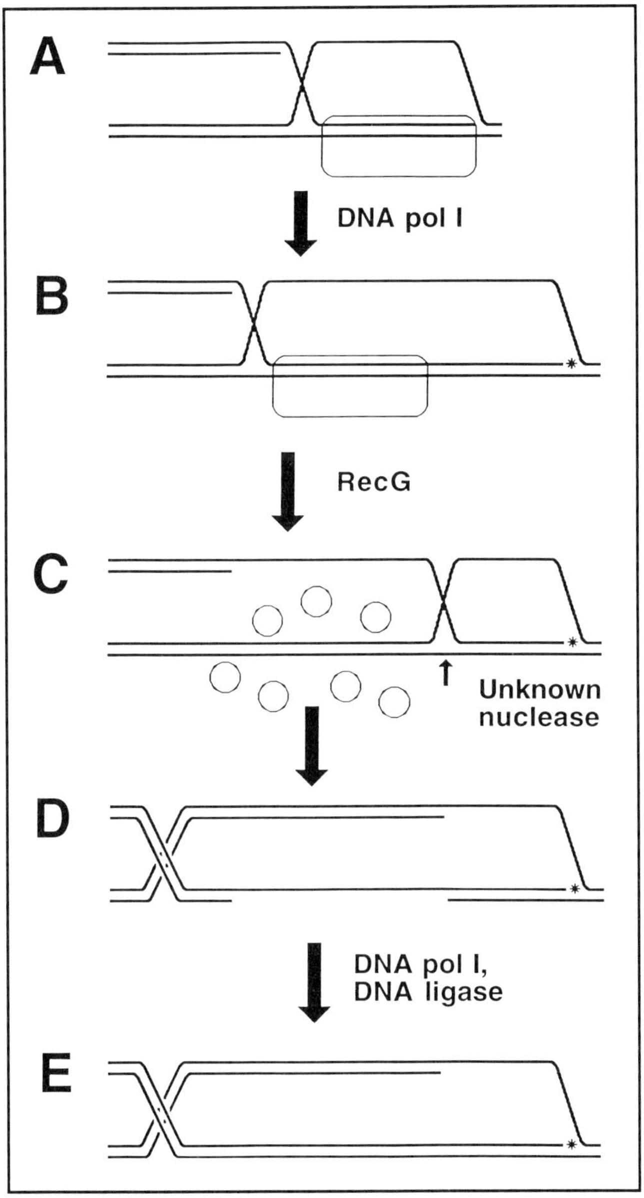

Fig. 5.8. An alternative scenario for the postsynaptic phase of double-strand end repair. RecA filament is shown as an open rectangle with rounded corners; RecA monomers (in "C") are shown as open circles. Limited DNA synthesis by DNA polymerase I is shown as a little star. Small arrow indicates the position of the required single-strand scission. Explanations in the text.

junctions, pushing them in the direction of growth of the RecA filament. If we assume that the invading strand ends 3' (which is the conventional assumption nowadays), then this directionality of RecG translocation coincides with the one required to remove the junction via the alternative mechanism (Fig. 5.8).

Thus, there could be two basic ways of removing junctions after recombinational repair of a disintegrated replication fork. When a junction is still three-stranded, it can be removed by incision of a single strand, either at the D-loop, or at the branching point. Three-stranded junctions are translocated by RecG, which should disperse used RecA filaments and might also help in the incision step (Fig. 5.8). An alternative resolution scenario might be called into action if a primosome assembly site is available on the displaced strand to initiate regular DNA replication. Such reinitiation will probably convert the three-strand junction into a Holliday junction, which will be used by RuvAB to disperse RecA filament and finally will be resolved by RuvC (Fig. 5.5D).

What conditions could favor switching from the RecG pathway to the RuvABC pathway? The RuvABC pathway seems to be specifically adapted to SOS-induced conditions, which arise whenever DNA synthesis is inhibited, and probably include elevated concentrations of Mg^{2+} and ATP (3.2). Under these conditions, RecA filaments may be more stable and difficult to remove. One way to think about the switching from the RecG to the RuvABC pathway is to assume that RecG can displace RecA filaments only when DNA replication is not grossly inhibited. Whenever inhibition of DNA replication causes accumulation of Mg^{2+} and ATP and appearance of dATP, RecG becomes inadequate against increasingly stable RecA filaments and has to give way to the RuvABC pathway, helping first to convert three-stranded junctions into Holliday junctions.[98] The proposed "division of labor" between RecG and RuvAB is consistent with the optimal Mg^{2+} concentrations for these two enzymes in vitro. Whereas the RuvAB optimum lies around 15 mM Mg^{2+} and coincides with the RecA optimum,[99] RecG works best at more "physiological" 5 mM concentrations of this cation.[94]

5.4.5. RecN and RorB – Still No Roles Assigned

In *E. coli*, *recBC* mutants are sensitive to both ionizing radiation and UV, which is understood as their inability to repair both double-strand breaks and disintegrated replication forks.[11] By contrast, yeast mutants sensitive to ionizing radiation due to deficiency in repair of double-strand breaks show little sensitivity to UV irradiation (reviewed in ref. 100). In *E. coli*, mutations combining intermediate sensitivity to ionizing radiation with almost WT resistance to UV radiation are also known. They are mapped to three separate loci called *rorA*, *rorB* and *radB* (*recN*).

rorA mutant is an allele of *recB*.[101] In vitro, *rorA*-RecBCD enzyme consumes more ATP during DNA degradation,[102] which may be the reason for the 2- to 3-fold increase in chromosomal degradation after ionizing irradiation observed in *rorA* mutants.[103] The increased DNA degradation should compromise the repair of double-strand breaks in the replicated portion of the chromosome leading to complete chromosome degradation if both daughter branches are broken in nearby positions (Fig. 5.9). This idea not only accounts for the *rorA* mutants' increased sensitivity to ionizing radiation, but is also compatible with their close to WT UV-resistance. Although the *rorA*-RecBCD enzyme should be also defective in repair of disintegrated replication forks (probably, the major lesion repaired by the RecBCD pathway after UV-irradiation), it still can simply degrade the resulting linear tail without killing the cell (5.5).

A possible nature of *rorA* mutation is revealed by the fact that 60-fold overproduction of the RecD subunit causes a similar phenotype: wild-type UV-resistance, decreased gamma-ray resistance and increased gamma-ray-induced DNA degradation.[76] A common belief that the removal of the RecD subunit turns the remaining enzyme into a recombinase, while its reassociation with the RecBC enzyme turns it into a degradase again (reviewed in ref. 57), accounts nicely for the phenotype of cells, overproducing the RecD subunit. Since the previous analysis makes it unlikely that the properties of *rorA*-RecBCD enzyme can be simply explained by the RecD polypeptide overproduction,[101,102] it might be that the mutant enzyme is less likely to let go of the RecD subunit when it encounters a Chi-site.

rorB mutants show intermediate sensitivity to gamma-radiation and cross-linking agents.[104] Their defect appears to be at later stages of double-strand break repair, as they show almost WT ability to rejoin the fragmented chromosome. The RorB protein was identified as a small 16.5 kDa protein with in vivo pI between 6.0 and 6.5.[105] In contrast, the sequence of the gene (p136 in ref. 106) agreeing on the size of the protein, predicts pI of 10.5, which suggests a posttranslational modification. No biochemical study of the RorB protein has been reported yet.

Viability of *E. coli* after exposure to ionizing radiation or mitomycin C (which cross-links DNA) is severely compromised in *radB*/*recN* mutants.[107-110] *recN* does not seem to be required for chromosomal rejoining after low doses of irradiation, but becomes increasingly important at higher doses of irradiation.[26] Normally, expression of *recN* is undetectable, however, it is greatly induced during the SOS response.[111] *recN* promoter has two binding sites for LexA repressor, which makes *recN* one of the most tightly-regulated genes of the SOS regulon.[112] The biochemical characterization of the RecN protein is yet to be done. Sequence analysis reveals an ATP-binding domain, but

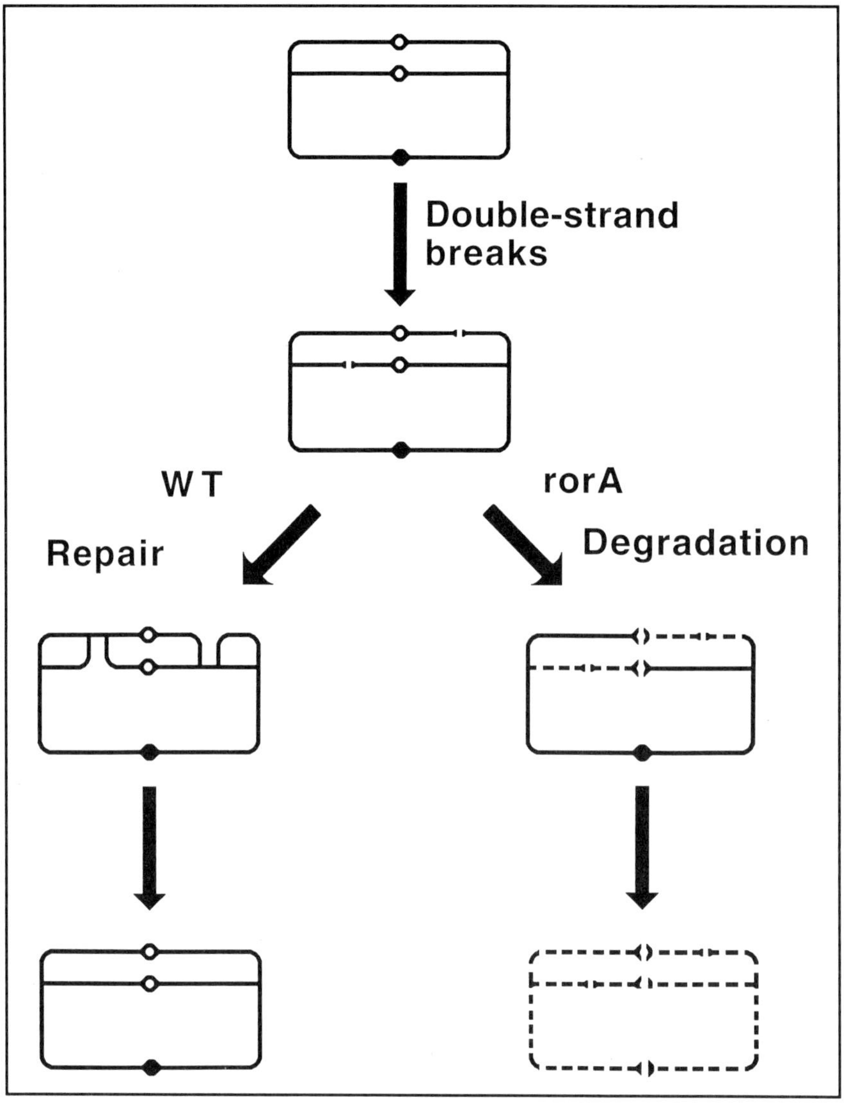

Fig. 5.9. The lethality caused by double-strand breaks in the replicated portion of the genome in rorA *strains. The chromosome is depicted as a theta-replicating structure. The origin of DNA replication is shown as a small open circle on the top of the chromosome, the terminus is shown as a closed circle at the bottom. If both daughter branches of a replicating chromosome are broken at positions close to each other, the double-strand breaks have more chances to be repaired in WT cells than in* rorA *cells, due to the increased DNA degradation in the latter.*

no other peculiarities that would hint to a possible function of RecN.[112] Since the initial rejoining of double-strand ends in *recN* mutants seems to be normal, while the mass rejoining efforts are blocked, RecN might be needed for "recharging" the early enzymes of the double-strand break-repair reaction (for example, for RecA turnover). This function of RecN is unlikely to be RecBCD-specific, as *recN* mutants are also defective or deficient in recombinational repair along RecBCD-independent pathways (5.6). It might be pertinent to note that degradation of damaged DNA in *recN* mutants is normal.[107]

5.5. DOUBLE-STRAND END REPAIR BACK-UP: DEGRADATION IN THE ABSENCE OF RECOMBINATIONAL REPAIR

Eighty percent of Chi-sites in the *E. coli* chromosome are oriented so as to stop RecBCD-degradation coming toward the replication origin.[113-117] This preferential orientation of Chi-sites suggests that the RecBCD repair pathway primarily deals with disintegrated replication forks,[11] probably because this type of lesion is more frequent in the *E. coli* chromosome than regular double-strand breaks (5.1.2).

While an unrepaired double-strand break in a chromosome is an obvious threat, it might be unclear why a disintegrated replication fork could not be simply ignored. However, if one of the two replication forks in a replication bubble has disintegrated while the other fork continues to function, chromosome overreplication will occur. In the case of circular bacterial chromosomes, this overreplication will be due to the switch of chromosomal replication from theta-mode to sigma-mode (Fig. 5.10B). In a circular chromosome with a single replication fork, initiation of a new round of theta-replication does not provide an exit from sigma-replication, but only increases the length of the linear tail (a sigma-replication trap) (Fig. 5.10B, E, F).

As argued above, in *E. coli*, after disintegration of a replication fork, the double-strand end is believed to be reattached to the circular chromosomal domain by RecA- and RecBCD-dependent repair (Fig. 5.10C). Therefore, the viability of strains which experience frequent replication fork disintegration should be dependent on both RecA and RecBCD enzymes. This is true for the most cases.[11] However, there are two *E. coli* mutants, *rnh* and *rep*, which are exceptions.

In the absence of RNase H (gp *rnh*), an enzyme degrading the RNA strand in RNA:DNA hybrids, replication of the *E. coli* chromosome initiates at several secondary origins of replications, *oriKs*, as well as from the main origin, *oriC*.[118,119] Some of the *oriKs* are within the terminus of the chromosome, so the progress of replication bubbles initiated at them is shortly blocked at the several termination sites bracketing the region.[120] Based on various evidence, it is suggested that stalled replication forks are eventually broken.[13,121] Interestingly, the viability of a strain with RNaseH inactivated depends mostly on ExoV

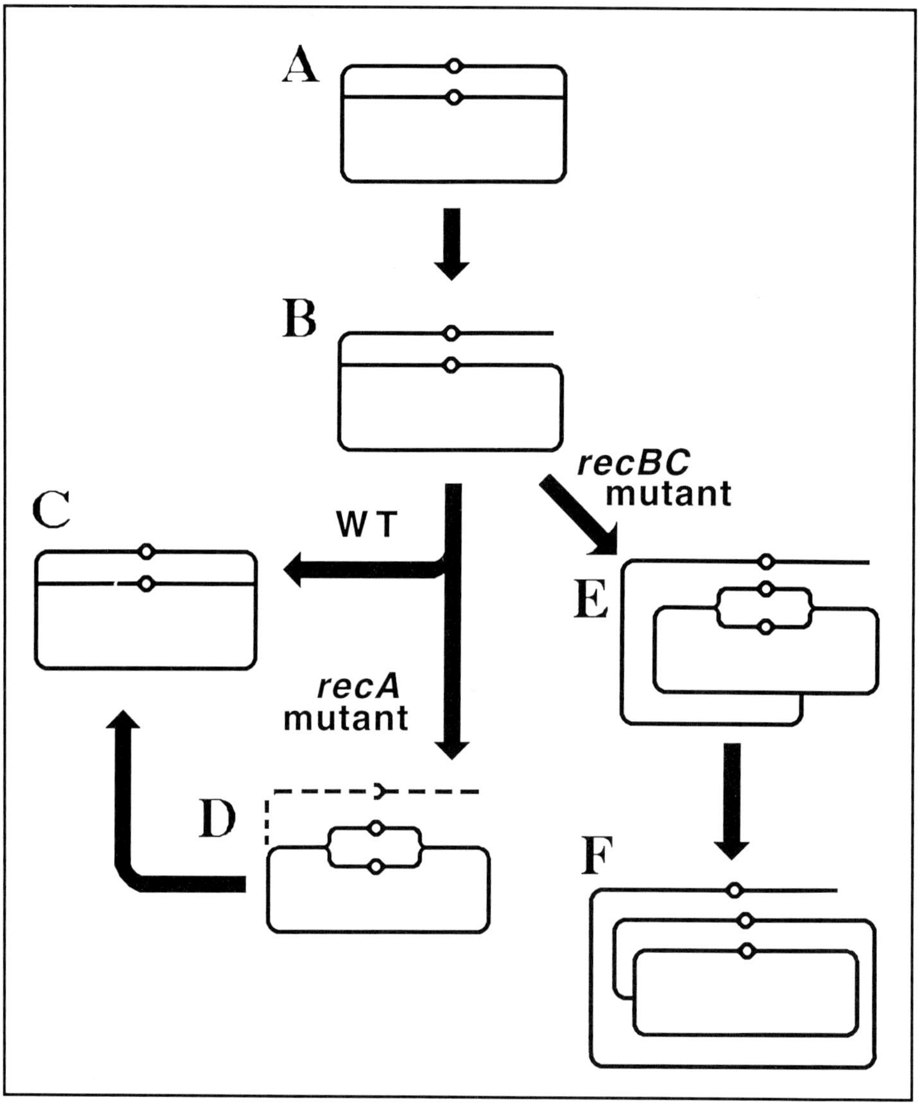

Fig. 5.10. An outline of circular chromosome maintenance. A theta-replicating chromosome (A) may suffer a collapse of one of the replication forks, becoming a sigma-replicating chromosome (B). In a wild-type strain, the double-stranded end is then degraded by RecBCD and reattached to the circular domain with the help of RecA (C). In the absence of RecA, the linear replicating branch is completely degraded by RecBCD, while a new round of theta-replication is initiated from the origin (D). In the absence of RecBCD, the chromosome is trapped in rolling-circle replication; reinitiation of theta-replication from the origin can only lengthen the linear tail (E, F). A similar scheme is proposed by Michel and associates.[124]

(DNA degradation), and is less dependent on RecBC or RecA (recombinational repair).[122]

rep mutants lack an auxiliary replication helicase, which makes their replication fork propagate with only half the speed of WT replication forks.[123] *rep recA* double mutants are barely viable, *rep recBC* mutants are not.[124] The authors interpret their results in terms of frequent disintegration of inhibited replication forks and propose that degradation of linear tails can maintain the normal replication pattern of the chromosome if recombinational repair fails.[124]

The following observations seem to reinforce this conclusion. *recD* mutants are recombination-proficient, still have 50% of the WT in vivo *recBC*-dependent exonuclease activity and survive DNA damage as well as or even better than the wild type cells when their growth is slow (5.3.1). However, in conditions allowing rapid growth a *recD* mutation may negatively affect survival after high doses of DNA damaging treatments.[125] RecA protein in these cells could be tied up in repair of lesions other than collapsed replication forks (for example, daughter-strand gaps). A back-up strategy for such an eventuality would be to degrade the linear branch all the way to the circular domain, eliminating the tail of a rolling-circle and returning the chromosome to theta-replication (Fig. 5.10D). *recD* cells, compromised in the degradational backup, would be able neither to repair the failed replication branches nor to make full use of the degradation option.

The back-up degradation is likely to be the responsibility of ExoV. In vitro, when presented with circular DNA molecules having linear tails, RecBCD degrades the tails but does not harm the circular domains,[126] demonstrating the feasibility of a similar reaction in vivo. The idea that the RecBCD degradation is a back-up for the reinstallation of disintegrated replication forks explains the differences in viabilities (a fraction of cells in population that are able to form colonies) in *recBC* versus *recA* mutant cultures. Both classes of mutants are deficient in fixing disintegrated replication forks; however, the viability of *recA* mutants is 60-70%, while the viability of *recBC* mutants is only 30%.[34] Although *recA* cells are deficient in recombinational repair, they can still degrade the failed replication branches with ExoV. On the other hand, *recBC* mutations are deficient not only in recombinational repair of disintegrated replication forks, but also in ExoV, so they may have no way out of the sigma-replication trap (Fig. 5.10). The proposed function for ExoV in controlling chromosome over-replication accounts for the ubiquitous presence of RecBCD analogs in bacteria.[32,40,41]

5.6. DOUBLE-STRAND END REPAIR IN THE ABSENCE OF RECBCD ENZYME

As mentioned in 5.3.1, null *recB* or *recC* mutants are deficient in double-strand end repair and lack ExoV, the major exonuclease activity.

Their poor capacity for the double-strand end repair, however, is ameliorated by various suppressor mutations. The resulting strains fall into two categories depending on their general nuclease activity. One category acquires a new exonuclease activity, while the other category lacks, in addition to exoV, a couple of less prominent nucleases. In both cases, these apparently opposite changes restore almost wild type capacity for double-strand end repair, demonstrating that the universal principles of such repair could be realized in a variety of ways. The knowledge about the alternative pathways for double-strand end repair in *E. coli* is important not only for our understanding of DNA metabolism in this bacterium. Eukaryotic cells lack strong bacterial-type nucleases, so their pathways of recombinational repair of double-strand ends are likely to have more in common with alternative pathways in *E. coli*, than with its major, RecBCD pathway.

5.6.1. The RecE Pathway

Chromosomes of *E. coli* strains isolated from nature carry genomes of a certain class of bacteriophages. Since the best studied phage of this class is called lambda, all other similarly structured phages are called lambdoid. The feature that enables lambdoid phages to contribute to the evolution of the *E. coli* genome is their ability to insert their chromosomes into the bacterial chromosome and to become dormant prophages. When in the chromosome, prophage genomes are exposed to the forces that continuously shape and reshape the bacterial genome. Deletions of dispensable genetic material are very important for maintaining bacterial chromosome slimness. The various parts of silent prophage genomes can be deleted without immediate consequences for the cell, severely crippling the prophage itself. Partly deleted and inactivated prophage genomes are called cryptic prophages, and their carcasses, as shipwrecks, litter the chromosomal shores of certain *E. coli* strains.[127]

Although incapacitated as independent entities, these remnants are still able to provide *E. coli* with phage functions when transcription of their genes is activated by additional mutations. Thus, *sbcA* mutations activate expression of a part of the Rac prophage, as a result of which at least two phage proteins relevant to recombinational repair are produced. RecE enzyme is a duplex DNA-specific exonuclease which selectively degrades the 5'-ending strand, producing a duplex DNA with a 3'-overhang.[128] RecT protein promotes annealing of complementary single DNA strands and can catalyze three-strand branch migration.[129] Possible roles of these two activities in bacteriophage recombination will be discussed in detail in chapter 6. A likely role of RecE exonuclease in *E. coli* repair is to resect double-strand ends so that they would terminate with long 3'-overhangs. Mutating away RecE exonuclease abolishes recombinational repair proficiency of *recBC sbcA* mutants.[130] The role of RecT protein, which is also needed for the full recombinational repair proficiency, is unclear.

Since RecE was the first gene of the alternative pathway for double-strand end repair to be discovered, the pathway was named after it.[131] The recombinational repair of the chromosomal DNA along the RecE pathway is dependent on *recA*,[132] *recE*,[132] *recF*,[132] *recJ*,[133] *recO*,[134] *recR*,[135] *recT*[130] and *ruvC*[136] genes. Understanding of how most of these genes operate during daughter-strand gap repair (chapter 4) allows visualization of a similar course of events catalyzed by the RecE pathway of double-strand end repair.

RecE exonuclease can act on blunt or nearly blunt ends or on ends with long 3'-overhangs. It can also start degradation of ends with short 5'-overhangs, but is inhibited by long 5'-overhangs,[128] which should be present in half of the double-strand ends generated as a result of replication fork disintegration (5.1.2) (Fig. 5.11). The removal of long 5'-overhangs could be a function of RecJ protein, a single-strand DNA-specific exonuclease, which degrades DNA strands starting from 5' ends.[137] After an end with a long 5'-overhang is made blunt by RecJ, RecE can start its degradation to create a 3'-overhang, which is immediately complexed by SSB. When the resection of the 5'-ending strand unravels a primosome assembly site on the 3'-ending strand, RNA primers are laid down, and DNA synthesis is initiated on the single-strand overhang (Fig. 5.11). Since the overhang has a 3'-polarity, it cannot be duplicated to the very end, and so the end-most primer cannot be removed (Fig. 5.11). As was proposed for the daughter-strand gap repair, RecF protein binds to the 5' side of every RNA primer and is subsequently displaced during completion of the synthesis of the next Okazaki piece (4.2.4). Since the RecF molecule sitting on the end-most primer cannot be displaced, it will catalyze the binding of RecR, RecO and, eventually, RecA, to the single-strand overhang (Fig. 5.11). The rest of the reaction is probably standard (Figs. 5.3 and 5.5). The weak effect of *ruvB*[136] and *recG*[91] mutations on recombinational repair along the RecE pathway can be understood in terms of redundant functions[90] (5.4.2).

5.6.2. The RecF Pathway

The RecF pathway of recombinational repair is turned on in *recBC* cells by inactivation of two additional nucleases. One nuclease that has to go is ExoI, which degrades single-strand DNA starting from 3' ends.[138] The gene coding for ExoI is called *sbcB*; ExoI is thought to participate in methyl-directed mismatch repair.[139] Another nuclease to be inactivated is SbcCD enzyme,[140,141] which is an ATP-dependent double-stranded DNA exonuclease (unpublished observations cited in ref. 142), distantly related to RecBCD.[143] It is not known why to turn on the RecF pathway one needs to inactivate ExoI and SbcCD. One

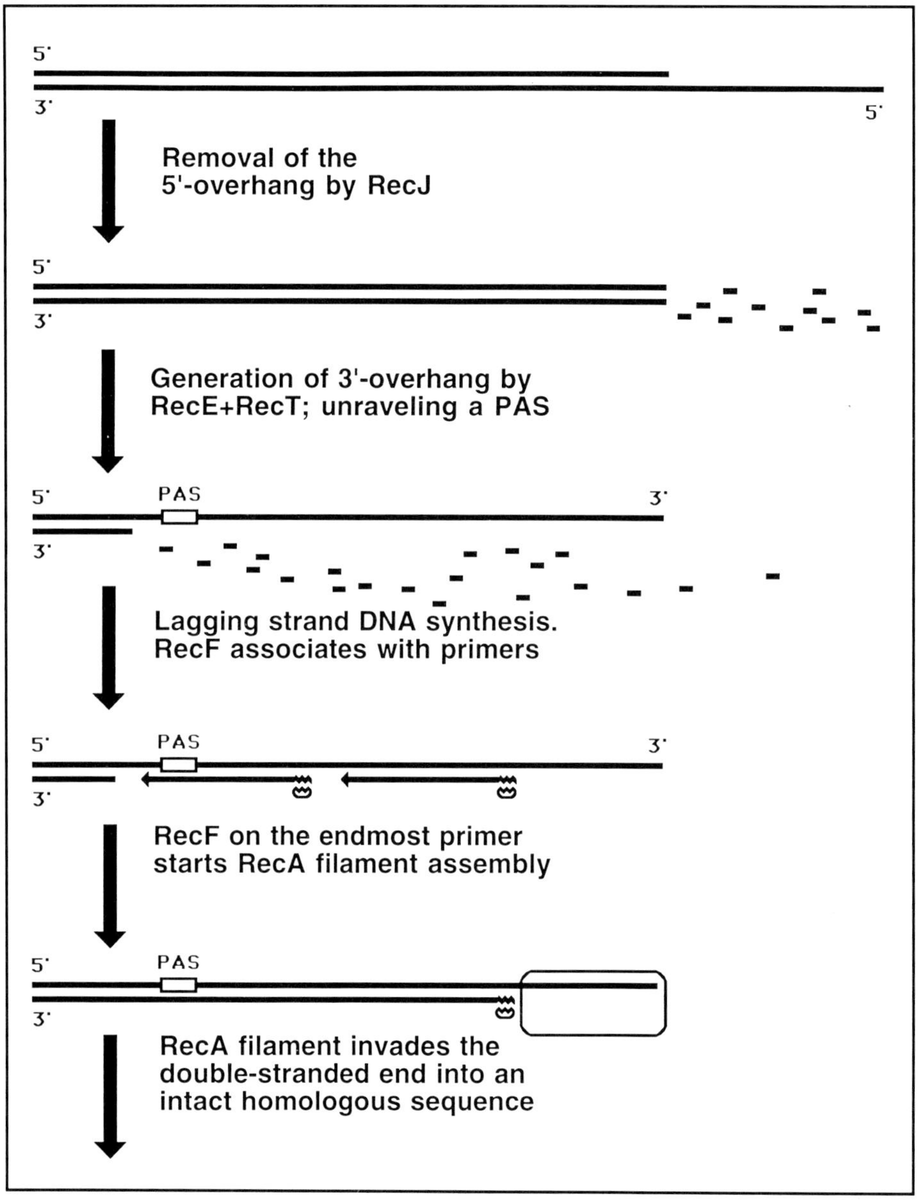

Fig. 5.11. A scheme for the presynaptic phase of the RecE pathway for double-strand end repair. PAS – primosome assembly site; short wavy segments with the associated little things – RNA primers with RecF protein; open rectangle with rounded corners – RecA filament. Explanation in the figure and in the text.

idea is that in the absence of ExoV the nuclease activity for generating 3'-overhangs is so weak that other nucleases that might degrade 3'-overhangs also need to be inactivated.[131] This idea assumes that only 3'-overhangs are active in RecA-mediated double-strand end invasion in vivo, which is still not shown conclusively.

The genetic requirements of the RecF recombinational repair pathway operating on the chromosome are similar to those of the RecE pathway: RecA,[144] RecF,[144] RecJ,[133] RecO,[134] RecR[135] and RuvC[96] proteins are needed. Besides, RecN,[108,134] RecQ,[145,146] RuvA and RuvB,[89] UvrD and HelD[146] proteins also participate in this pathway. The requirements for RecQ, UvrD and HelD (all three are DNA helicases with similar behavior in vitro (reviewed in ref. 147)) can be accounted for by alternative ways of generating 3'-overhangs. In the RecF pathway, 3'-overhangs could be produced by combined action of either RecQ helicase (or UvrD/HelD helicases) and RecJ exonuclease[137,148] (Fig. 5.12). Everything else could be as in the RecE pathway (explaining their mostly overlapping requirements). The only protein whose function remains unexplained by this scheme is RecN.

5.6.3. A Unified Mechanism for Double-Strand End Repair

The discussed mechanisms for the reattachment of double-strand ends along the RecBCD, the RecE or the RecF pathways have a lot in common. In fact, their distinction is only in the presynaptic reactions, which, although structurally similar, are catalyzed by different enzymes. Therefore, one can envisage a unified mechanism for the double-strand end repair (Fig. 5.12).[148]

The first stage is processing of a double-strand end to generate a 3'-single-strand overhang. RecA-filament is then assembled on the single-strand overhang to catalyze invasion of the end into an intact duplex. A three-strand junction is generated, and a limited DNA synthesis is primed by DNA pol I at the invading 3'-end. From this point, there is likely to exist two alternative ways of completing the double-strand end repair, each one ensuring the displacement of the used RecA filament and removal of the junction. One pathway is controlled by RecG helicase, which translocates the three-way junction until it is removed by a single-strand scission. The alternative, RuvABC-dependent pathway, may operate only when the three-strand junction is converted by DNA synthesis into a Holliday junction.

5.7. DEMULTIMERIZATION OF THE CHROMOSOME AFTER DOUBLE-STRAND END REPAIR

When a double-strand end invades a homologous duplex DNA, a single junction is formed, which is either resolved by RuvC or removed in the RecG-dependent pathway. If it is a Holliday junction and is still covered by RecA filament, it has a specific configuration

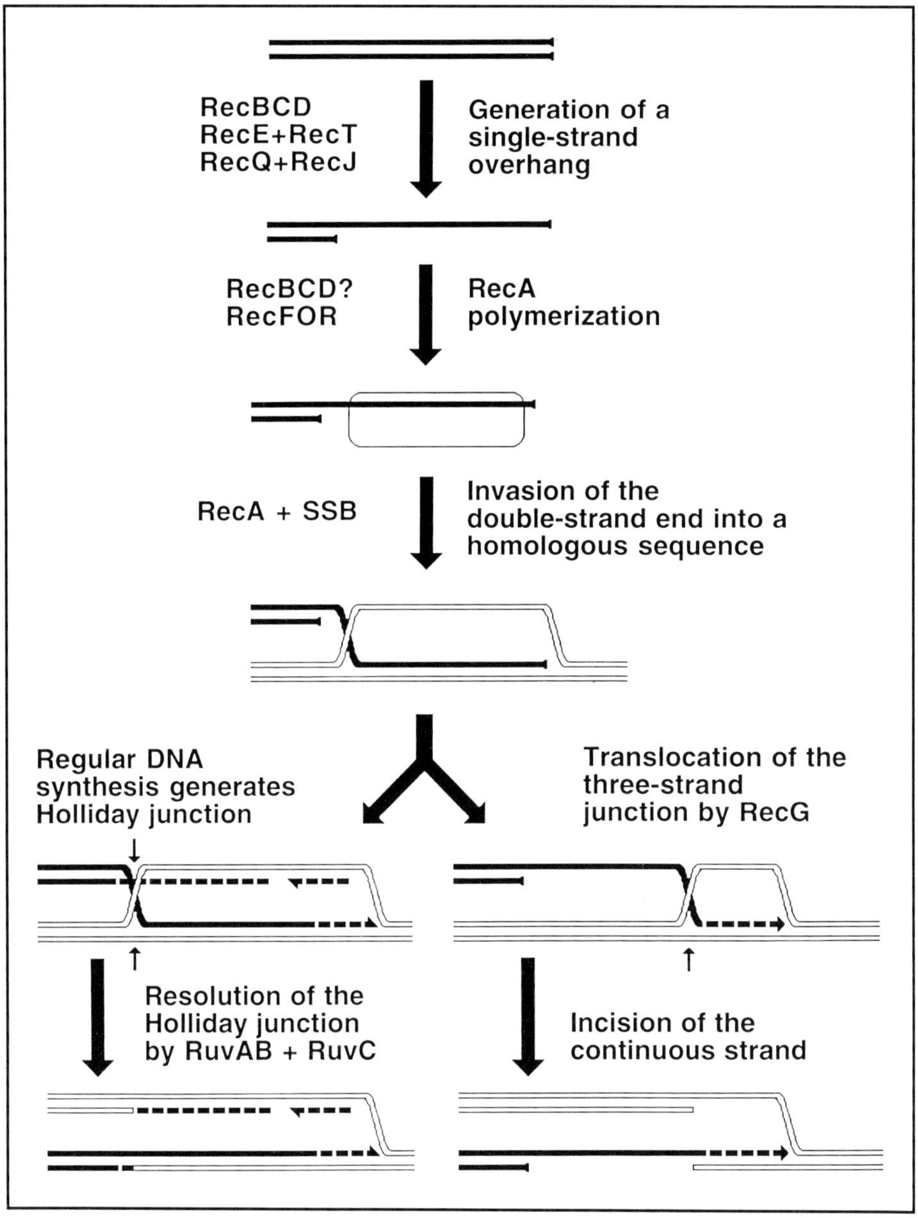

Fig. 5.12. A unified scheme for double-strand end repair. RecA filament is shown as an open rectangle with rounded corners, positions of single-strand scissions are indicated by small arrows. Explanations in the figure and in the text.

that dictates the direction in which it will be resolved by RuvC. In vitro, RuvC-catalyzed resolution of RecA-covered Holliday junctions ensures the appearance of a cross-over behind the restored replication fork[22,149] (Fig. 5.13A). A three-strand junction can be removed only with a concomitant generation of a cross-over (Fig. 5.13A). In other words, if there is a single DSE repair event per chromosomal replication, the resulting sister genomes are likely to find themselves combined into a single dimeric chromosome (Fig. 5.13B).

The unique susceptibility of circular chromosomes to odd numbers of exchanges was first recognized by McClintock in her studies with maize ring chromosomes.[150] However, eukaryotes do not usually face this problem, as all the known natural eukaryotic chromosomes are linear. In contrast, prokaryotes, with few exceptions, have circular chromosomes, so they need a system to monomerize the chromosomes that became dimeric as a result of recombinational repair. Such a system, based on site-specific recombination, has been recently elucidated in *E. coli*.

5.7.1. Site-Specific Demultimerization of the Chromosome in the Terminus Region: XerC and XerD Dependent Reaction at *dif* Site

E. coli chromosome has a single replication origin generating two divergent replication forks which move around the chromosome with approximately equal speed. Replication forks converge in the region situated on the chromosome across from the replication origin, called the replication terminus (1.3). In the middle of the terminus, there is a 30 bp site called *dif* which is responsible for the chromosome monomerization.[151-153] A pair of enzymes coded by unlinked *xerC* and *xerD* genes are found to work on this site to demultimerize the chromosomes[151,154] (Fig. 5.14). Mutating away *xerC* or deleting *dif* site causes cells filamentation due to problems with nucleoid partitioning. Mutating away *recA* or *recB* genes alleviates these problems of *xerC* or *dif* mutants,[151,153] which points to the RecABCD-catalyzed repair of disintegrated replication forks as a cause of nucleoid partitioning problems. It is proposed that at least 20% of the cells experience an inducing event (most likely, replication fork collapse and repair) which subsequently leads to partitioning problems in *dif* mutants.[153] In a good agreement with this assessment, about 25% of *recA* cells (which degrade the linear tails instead of repairing the collapsed replication forks (5.5)) have at least one nucleoid missing.[155]

The XerCD-mediated reaction is studied in vivo with plasmid substrates bearing the 30 bp *dif* sequence.[151-154] In this setting, the reaction shows no resolution selectivity: two *dif* sites are recombined independently of whether they are on a single DNA molecule or on two separate DNA molecules. The lack of selectivity poses a potential problem, since it is unclear how the XerCD-*dif* system is able to faithfully resolve

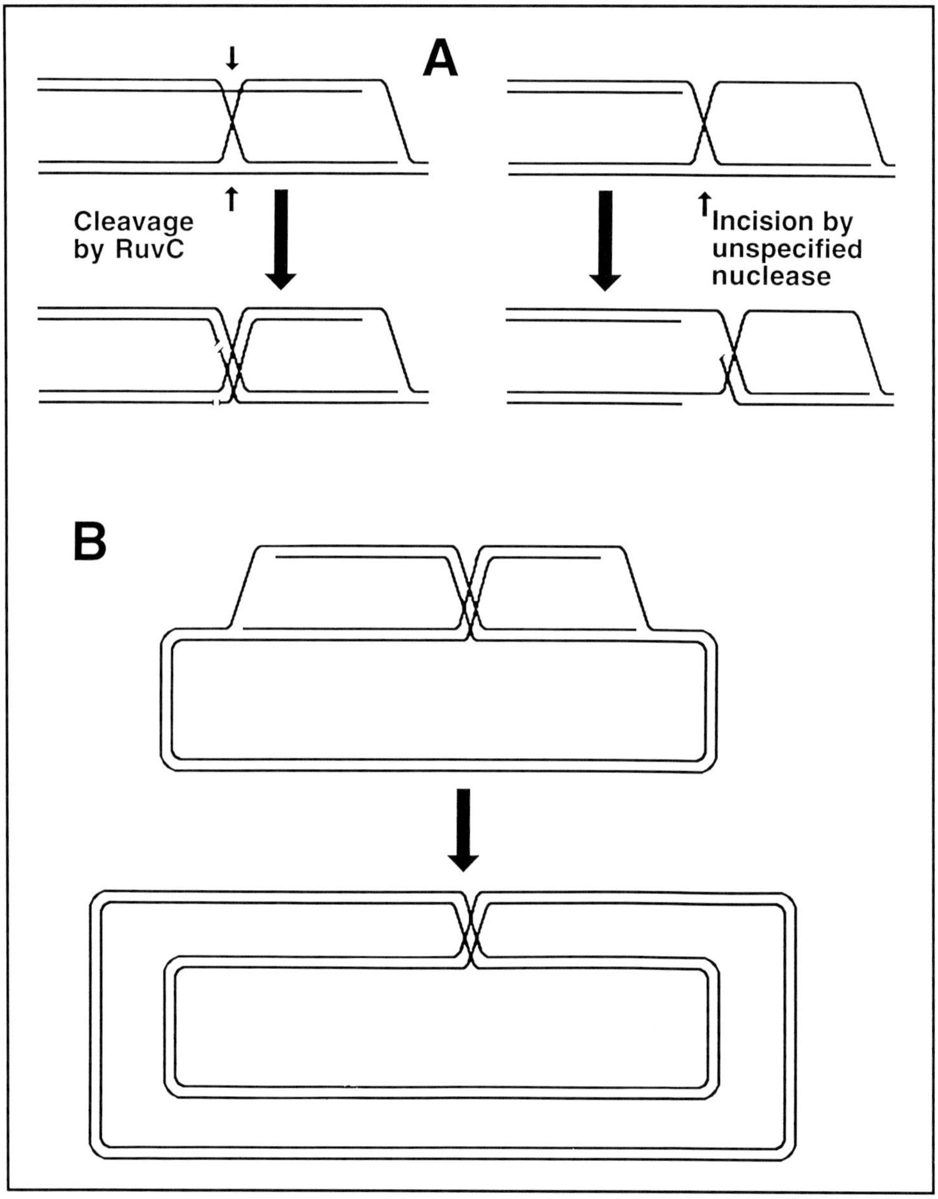

Fig. 5.13. Removal of the single junction leads to a cross-over and to dimerization of a circular chromosome. (A) Crossover is formed behind a restored replication fork. Positions of single-strand scissions are indicated by small arrows. (B) A single cross-over in a circular chromosome translates into a dimer chromosome.

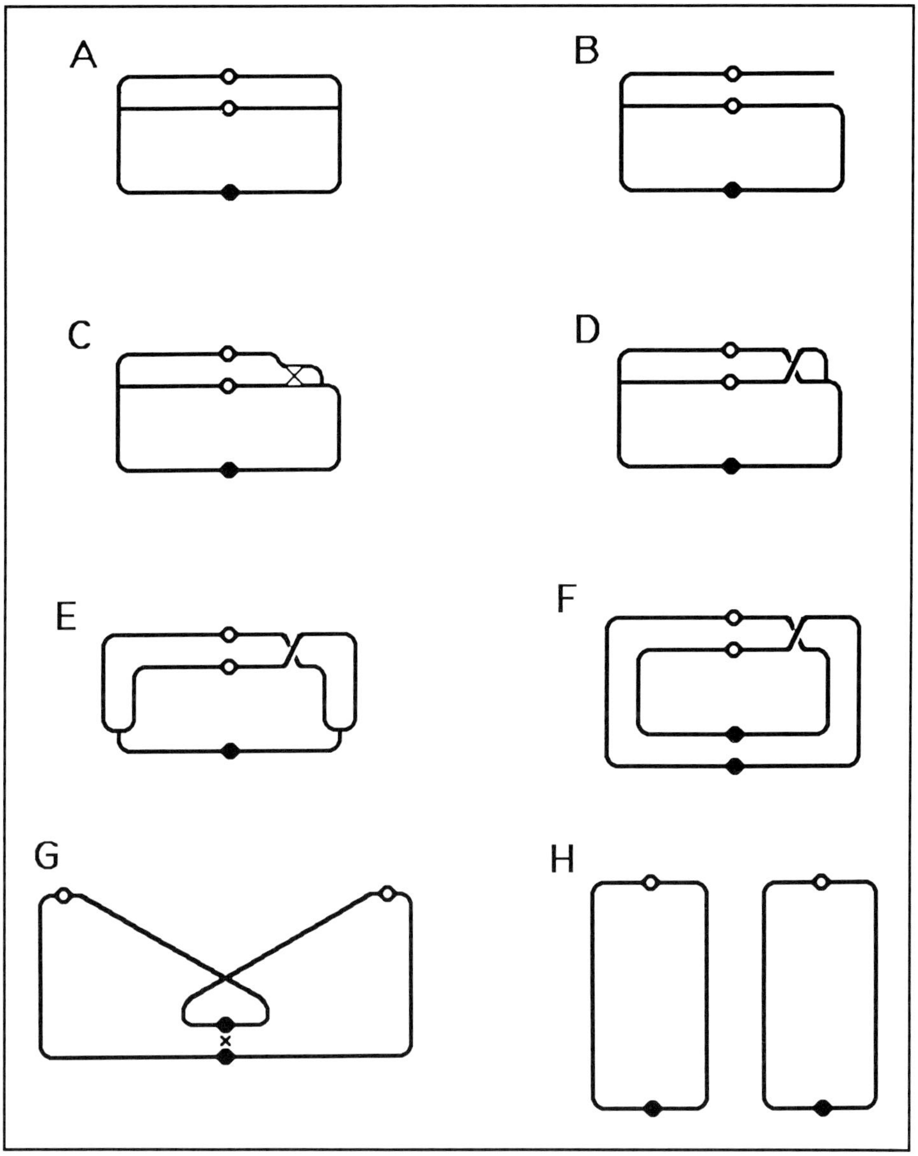

Fig. 5.14. XerCD-dif *site specific recombination ensures chromosome demultimerization in* E. coli. *The chromosome is shown as a rectangle with rounded corners; replication origin is depicted as a small open circle on the top of this rectangle, the replication terminus is represented as a small filled circle at the bottom of the chromosome. (A) A chromosome replicating in theta-mode. (B) One of the replication forks has disintegrated. (C) The replication fork is reassembled back by recombinational repair resulting in a Holliday junction. (D) The Holliday junction is resolved resulting in a cross-over. (E and F) The complete replication of the resulting structure produces a dimer chromosome. (G) Segregation of the daughter nucleoids drives the cross-over to the terminus region, where it is resolved, separating the daughter chromosomes from each other (H).*

chromosomal dimers. It is postulated that the XerCD-*dif* system is able to demultimerize chromosomes due to frequent exchanges between two *dif* sites on sister chromosomes and active separation of the sister chromosomes by partitioning apparatus of the cell.[151,153] Alternatively, one can imagine that there is a supramolecular structure of the DNA region surrounding *dif* site which renders the XerCD-mediated reaction unidirectional. *loxP*/Cre site-specific recombinational system of bacteriophage P1 substitutes for *dif* in chromosomal demultimerization if inserted in the terminus region.[156] Neither *dif*/XerCD, nor *loxP*/Cre are active in chromosome demultimerization when transplanted to a position close to the origin of DNA replication or halfway between the origin and the terminus. These results suggest that it is not the special site-specific recombination system, but the position of the resolution site within the terminus, that is important for chromosomal demultimerization in *E. coli*.[156]

5.7.2. Evolution of the *E. coli* Genome by Two Duplications

Many genes coding for proteins with related metabolic functions or with significant homologies are separated on the *E. coli* chromosome by a distance which is either 1/4 or 1/2 of the chromosomal length.[157] Thus, across the RecBCD gene cluster, there is *sbcCD* operon which is partly homologous to *recB* and *recC* genes[143] and codes for an enzyme of DNA metabolism which prevents propagation of chromosomes containing big palindromes.[141] Another illustration is a silent replication origin located in the chromosomal terminus, which becomes functional along with *oriC* in certain mutants.[119] This symmetrical disposition of related functions is taken to mean that contemporary *E. coli* chromosome evolved as a result of two consecutive duplications of the primordial genome.[157] These duplications could happen, for example, when XerCD-*dif* system failed to resolve a dimer chromosome. To be fixed, such duplications must have been followed by an array of smaller rearrangements and mutations, modifying certain duplicated functions like replication origins or termination sites. These secondary rearrangements would smear the exact symmetry of the original duplicated chromosome and sometimes separate genes and sites that were originally located close to each other. For example, in the majority of site-specific recombination systems, the recombinase gene is located right near its site of action. In contrast, *xerC*, *xerD* and *dif* are widely separated, *xerC* being approximately 1/4 of the chromosome away from *xerD* and 1/2 of the chromosome away from *dif*. There might have been originally a single locus on the primordial chromosome responsible for the genome demultimerization, which, as a result of two genome duplications and subsequent deletions and mutations have evolved into several separate loci performing the same function in the contemporary *E. coli* genome.

SUMMARY

Double-strand breaks in chromosomal DNA are induced in vivo by ionizing radiation or during some enzymatic reactions. The two ends of a double-strand break are probably repaired independently of each other, justifying the term "double-strand end repair". A solitary double-strand end is generated as a result of a replication fork disintegration. In the presynaptic phase of double-strand end repair, helicases and nucleases make the end single-stranded to allow RecA polymerization. After the RecA-catalyzed invasion of the end into an intact homologous duplex, a replication fork is formed and the junction is removed.

In *E. coli*, the WT pathway of double-strand end repair is catalyzed by RecA protein with the help of RecBCD enzyme. RecBCD is an extremely potent helicase which periodically clips off the unwound strands, thus degrading DNA from a double-strand end. RecBCD transiently stops its degradation spree at Chi-sites, probably waiting for RecA protein to begin its filament assembly on the single-strand loop, held by RecBCD. RecA polymerization, besides starting homologous search, could also inactivate the nuclease activity of RecBCD. After a successful invasion of the frayed end into an intact duplex, a three-stranded junction is removed by the RecG pathway of postsynaptic repair, while a Holliday junction (if formed) is removed by the RuvABC pathway.

If RecA is in short supply, RecBCD is likely to resume its degradative movement past Chi. According to this idea, RecBCD will be able to degrade the linear branch as long as RecA is unavailable for the double-strand end invasion. DNA degradation is a back-up that prevents uncontrolled chromosomal replication if a disintegrated replication fork cannot be restored. In the absence of RecBCD enzyme, two alternative pathways for double-strand end repair can be activated by either expression or suppression of additional nucleases. In both cases, the repair proficiency is associated with the re-acquired capacity to generate double-strand ends with 3'-single-strand overhangs.

Repair of a disintegrated replication fork in the circular *E. coli* chromosome often leaves behind a dimeric chromosome. A site-specific recombination system in the terminus makes sure the dimer is resolved during daughter nucleoid segregation prior to cell division. The current *E. coli* chromosome might be a result of at least two ancient dimerizations.

Repair of disintegrated replication forks and double-strand breaks in other bacteria is likely to follow in major details the *E. coli* scheme. ATP-dependent nucleases (analogous to RecBCD enzyme) are found in many bacteria,[41,158] and Chi-sites are characterized for at least one more microbe.[63] RecG homolog is found in *Haemophilus influenzae*;[159] XerCD site-specific recombination system is present in several enterobacteria.[160]

The enzymatic complexity of the double-strand break repair in the *E. coli* chromosome is impressive. The next chapter discusses a simpler alternative, employed by some bacteriophages, whose genomes are only 1-2% of the *E. coli* genome. The phage way of repairing double-strand breaks is also recombinational and therefore faithful, but how enzymatically simple and elegant it is compared to the host DSB repair! However, saving on enzymes, phages have to be wasteful on their DNA, which is quite acceptable for them but would be unacceptable for the *E. coli* chromosome.

References

1. Krasin F and Hutchinson F (1977) Repair of DNA double-strand breaks in *Escherichia coli*, which requires *recA* function and the presence of duplicate genome. J Mol Biol 116:81-98.
2. Krisch RE, Krasin F and Sauri CJ (1976) DNA breakage, repair and lethality after ^{125}I decay in rec$^+$ and *recA* strains of *Escherichia coli*. Int J Radiat Biol 29:37-50.
3. Ulmer KM, Gomez RF and Sinskey AJ (1979) Ionizing radiation damage to the folded chromosome of *Escherichia coli* K-12: repair of double-strand breaks in deoxyribonucleic acid. J Bacteriol 138:486-491.
4. Bonura T, Town CD, Smith KC and Kaplan HS (1975) The influence of oxygen on the yield of DNA double-strand breaks in X-irradiated *Escherichia coli* K-12. Radiat Res 63:567-577.
5. Bender J and Kleckner N (1986) Genetic evidence that Tn*10* transposes by a nonreplicative mechanism. Cell 45:801-815.
6. Hagemann AT and Craig NL (1993) Tn*7* transposition creates a hotspot for homologous recombination at the transposon donor site. Genetics 133:9-16.
7. Chang S and Cohen SN (1977) In vivo site-specific genetic recombination promoted by the *Eco*RI restriction endonuclease. Proc Natl Acad Sci USA 74:4811-4815.
8. Naito T, Kusano K and Kobayashi I (1995) Selfish behavior of restriction-modification systems. Science 267:897-899.
9. Szostak JW, Orr-Weaver TL, Rothstein RJ and Stahl FW (1983) The double-strand-break repair model for recombination. Cell 33:25-35.
10. Resnick MA (1976) The repair of double-strand breaks in DNA: a model involving recombination. J Theor Biol 59:97-106.
11. Kuzminov A (1995) Collapse and repair of replication forks in *Escherichia coli*. Mol Microbiol 16:373-384.
12. Kuzminov A (1995) Instability of inhibited replication forks in *E. coli*. BioEssays 17:733-741.
13. Horiuchi T and Fujimura Y (1995) Recombinational rescue of the stalled DNA replication fork: a model based on analysis of an *Escherichia coli* strain with a chromosomal region difficult to replicate. J Bacteriol 177:783-791.

14. Sharma B and Hill TM (1995) Insertion of inverted *Ter* sites into the terminus region of the *Escherichia coli* chromosome delays completion of DNA replication and disrupts the cell cycle. Mol Microbiol 18:45-61.
15. Roth D and Wilson J (1988) Illegitimate recombination in mammalian cells. In: Kucherlapati R and Smith GR, eds. Genetic Recombination. Washington, D.C.: American Society for Microbiology, pp 621-653.
16. Jeggo PA, Taccioli GE and Jackson SP (1995) Menage à trois: double strand break repair, V(D)J recombination and DNA-PK. BioEssays 17:949-957.
17. Tsaneva IR and West SC (1994) Targeted *versus* non-targeted DNA helicase activity of the RuvA and RuvB proteins of *Escherichia coli.* J Biol Chem 269:26552-26558.
18. Register JC III and Griffith J (1985) The direction of RecA protein assembly onto single-strand DNA is the same as the direction of strand assimilation during strand exchange. J Biol Chem 260:12308-12312.
19. Siddiqi I, Stahl MM and Stahl FW (1991) Heteroduplex chain polarity in recombination of phage lambda by the red, RecBCD, RecBC(D^-) and RecF pathways. Genetics 128:7-22.
20. Hagemann AT and Rosenberg SM (1991) Chain bias in Chi-stimulated heteroduplex patches in the λ *ren* gene is determined by the orientation of λ *cos*. Genetics 129:611-621.
21. Asai T and Kogoma T (1994) D-loops and R-loops: alternative mechanisms for the initiation of chromosome replication in *Escherichia coli.* J Bacteriol 176:1807-1812.
22. Benson FE and West SC (1994) Substrate specificity of the *Escherichia coli* RuvC protein. Resolution of three- and four-stranded recombination intermediates. J Biol Chem 269:5195-5201.
23. Shah R, Bennett RJ and West SC (1994) Genetic recombination in *E. coli*: RuvC protein cleaves Holliday junctions at resolution hotspots in vitro. Cell 79:853-864.
24. Glassberg J, Meyer RR and Kornberg A (1979) Mutant single-strand binding protein of *Escherichia coli*: genetic and physiological characterization. J Bacteriol 140:14-19.
25. Ennis DG, Amundsen SK and Smith GR (1987) Genetic functions promoting homologous recombination in *Escherichia coli*: a study of inversion in phage λ. Genetics 115:11-24.
26. Sargentini NJ and Smith KC (1986) Quantitation of the involvement of the *recA, recB, recC, recF, recJ, recN, lexA, radA, radB, uvrD,* and *umuC* genes in the repair of X-ray-induced DNA double-strand breaks in *Escherichia coli.* Radiat Res 107:58-72.
27. Sasaki M, Fujiyoshi T, Shimada K and Takagi Y (1982) Fine structure of the *recB* and *recC* gene region of *Escherichia coli.* Biochem Biophys Res Commun 109:414-422.
28. Amundsen SK, Taylor AF, Chaudhury AM and Smith GR (1986) *recD*: the gene for an essential third subunit of exonuclease V. Proc Natl Acad Sci USA 83:5558-5562.

29. Finch PW, Storey A, Brown K, Hickson ID and Emmerson PT (1986) Complete nucleotide sequence of *recD*, the structural gene for the alpha subunit of exonuclease V of *Escherichia coli*. Nucleic Acids Res 14:8583-8594.
30. Finch PW, Storey A, Chapman KE, Brown K, Hickson ID and Emmerson PT (1986) Complete nucleotide sequence of the *Escherichia coli recB* gene. Nucleic Acids Res 14:8573-8582.
31. Finch PW, Wilson RE, Brown K, Hickson ID, Tomkinson AE and Emmerson PT (1986) Complete nucleotide sequence of the *Escherichia coli recC* gene and of the *thyA-recC* intergenic region. Nucleic Acids Res 14:4437-4451.
32. Taylor AF (1988) RecBCD enzyme of *Escherichia coli*. In: Kucherlapati R and Smith GR, eds. Genetic Recombination. Washington, D.C.: American Society for Microbiology, pp 231-263.
33. Willetts NS and Mount DW (1969) Genetic analysis of recombination-deficient mutants of *Escherichia coli* K-12 carrying *rec* mutations cotransducible with *thyA*. J Bacteriol 100:923-934.
34. Capaldo FN, Ramsey G and Barbour SD (1974) Analysis of the growth of recombination-deficient strains of *Escherichia coli* K-12. J Bacteriol 118:242-249.
35. Wang T-CV and Smith KC (1986) Postreplicational formation and repair of DNA double-strand breaks in UV-irradiated *Escherichia coli uvrB* cells. Mutat Res 165:39-44.
36. Chaudhury AM and Smith GR (1984) A new class of *Escherichia coli recBC* mutants: implications for the role of RecBC enzyme in homologous recombination. Proc Natl Acad Sci USA 81:7850-7854.
37. Rinken R, Thoms B and Wackernagel W (1992) Evidence that *recBC*-dependent degradation of duplex DNA in *Escherichia coli recD* mutants involves DNA unwinding. J Bacteriol 174:5424-5429.
38. Taylor AF and Smith GR (1995) Monomeric RecBCD enzyme binds and unwinds DNA. J Biol Chem 270:24451-24458.
39. Wright M, Buttin G and Hurwitz J (1971) The isolation and characterization from *Escherichia coli* of adenosine triphosphate-dependent deoxyribonuclease directed by *recB, C* genes. J Biol Chem 246:6543-6555.
40. Clark AJ (1973) Recombination deficient mutants of *E. coli* and other bacteria. Annu Rev Genet 7:67-86.
41. Telander-Muskavitch KM and Linn S (1981) RecBC-like enzymes: exonuclease V deoxyribonucleases. In: Boyer PD, ed. The Enzymes. New York: Academic Press, Inc., pp 234-250.
42. Goldmark PJ and Linn S (1972) Purification and properties of RecBC DNase of *Escherichia coli* K-12. J Biol Chem 247:1849-1860.
43. Karu AE, MacKay V, Goldmark PJ and Linn S (1973) The RecBC deoxyribonuclease of *Escherichia coli* K-12. Substrate specificity and reaction intermediates. J Biol Chem 248:4874-4884.
44. Eichler DC and Lehman IR (1977) On the role of ATP in phosphodiester bond hydrolysis catalyzed by the recBC deoxyribonuclease of *Escherichia coli*. J Biol Chem 252:499-503.

45. Taylor AF and Smith GR (1985) Substrate specificity of the DNA unwinding activity of the RecBC enzyme of *Escherichia coli.* J Mol Biol 185:431-443.
46. Garfinkel L and Garfinkel D (1984) Calculation of free-Mg^{2+} concentration in adenosine 5'-triphosphate containing solutions in vitro and in vivo. Biochemistry 23:3547-3552.
47. Eggleston AK and Kowalczykowski SC (1993) Biochemical characterization of a mutant RecBCD enzyme, the $RecB^{2109}CD$ enzyme, which lacks χ-specific, but not non-specific, nuclease activity. J Mol Biol 231:605-620.
48. Dixon DA and Kowalczykowski SC (1995) Role of the *Escherichia coli* recombination hotspot, χ, in RecABCD-dependent homologous pairing. J Biol Chem 270:16360-16370.
49. Roman L and Kowalczykowski SC (1989) Characterization of the helicase activity of the *Escherichia coli* RecBCD enzyme using a novel helicase assay. Biochemistry 28:2863-2873.
50. Roman LJ, Eggleston AK and Kowalczykowski SC (1992) Processivity of the DNA helicase activity of *Escherichia coli* RecBCD enzyme. J Biol Chem 267:4207-4214.
51. Julin DA and Lehman IR (1987) Photoaffinity labeling of the RecBCD enzyme of *Escherichia coli* with 8-azidoadenosine 5'-triphosphate. J Biol Chem 262:9044-9051.
52. Masterson C, Boehmer PE, McDonald F, Chaudhury S, Hickson ID and Emmerson PT (1992) Reconstitution of the activities of the RecBCD holoenzyme of *Escherichia coli* from the purified subunits. J Biol Chem 267:13564-13572.
53. Boehmer PE and Emmerson PT (1992) The RecB subunit of the *Escherichia coli* RecBCD enzyme couples ATP hydrolysis to DNA unwinding. J Biol Chem 267:4981-4987.
54. Dixon DA and Kowalczykowski SC (1993) The recombination hotspot χ is a regulatory sequence that acts by attenuating the nuclease activity of the *E. coli* RecBCD enzyme. Cell 73:87-96.
55. Ganesan S and Smith GR (1993) Strand-specific binding to duplex DNA ends by the subunits of the *Escherichia coli* RecBCD enzyme. J Mol Biol 229:67-78.
56. Taylor AF and Smith GR (1980) Unwinding and rewinding of DNA by the RecBC enzyme. Cell 22:447-457.
57. Myers RS and Stahl FW (1994) χ and the RecBCD enzyme of *Escherichia coli.* Annu Rev Genet 28:49-70.
58. Enquist LW and Skalka A (1973) Replication of bacteriophage λ DNA dependent on the function of host and viral genes. I. Interaction of *red, gam,* and *rec.* J Mol Biol 75:185-212.
59. Stahl FW, Thomason LC, Siddiqi I and Stahl MM (1990) Further test of a recombination model in which χ removes the RecD subunit from the RecBCD enzyme of *Escherichia coli.* Genetics 126:519-533.
60. Dabert P, Ehrlich SD and Gruss A (1992) χ sequence protects against RecBCD degradation of DNA in vivo. Proc Natl Acad Sci USA 89:12073-12077.

61. Kuzminov A, Schabtach E and Stahl FW (1994) χ-sites in combination with RecA protein increase the survival of linear DNA in *E. coli* by inactivating exoV activity of RecBCD nuclease. EMBO J 13:2764-2776.
62. Smith GR, Kunes SM, Schultz DW, Taylor A and Triman KL (1981) Structure of Chi hotspots of generalized recombination. Cell 24:429-436.
63. Biswas I, Maguin E, Ehrlich SD and Gruss A (1995) A 7-base-pair sequence protects DNA from exonucleolytic degradation in *Lactococcus lactis*. Proc Natl Acad Sci USA 92:2244-2248.
64. Faulds D, Dower N, Stahl MM and Stahl FW (1979) Orientation-dependent recombination hotspot activity in bacteriophage λ. J Mol Biol 131:681-695.
65. Yagil E, Dower NA, Chattoraj D, Stahl M, Pierson C and Stahl FW (1980) Chi mutation in a transposon and the orientation-dependence of Chi phenotype. Genetics 96:43-57.
66. Taylor AF, Schultz DW, Ponticelli AS and Smith GR (1985) RecBC enzyme nicking at Chi sites during DNA unwinding: location and orientation-dependence of the cutting. Cell 41:153-163.
67. Yagil E and Shtromas I (1985) Rec-mediated recombinational activity of two adjacent Chi elements in bacteriophage lambda. Genet Res 45:1-8.
68. Taylor AF and Smith GR (1992) RecBCD enzyme is altered upon cutting DNA at a Chi recombinational hotspot. Proc Natl Acad Sci USA 89:5226-5230.
69. Taylor AF and Smith GR (1995) Strand specificity of nicking of DNA at Chi sites by RecBCD enzyme. Modulation by ATP and magnesium levels. J Biol Chem 270:24459-24467.
70. Cheng KC and Smith GR (1987) Cutting of Chi-like sequences by the RecBCD enzyme of *Escherichia coli*. J Mol Biol 194:747-750.
71. Thaler DS, Sampson E, Siddiqi I, Rosenberg SM, Thomason LC, Stahl FW and Stahl MM (1989) Recombination of bacteriophage λ in *recD* mutants of *Escherichia coli*. Genome 31:53-67.
72. Thaler DS, Sampson E, Siddiqi I, Rosenberg SM, Stahl FW and Stahl MM (1988) A hypothesis: Chi-activation of RecBCD enzyme involves removal of the RecD subunit. In: Friedberg E and Hanawalt P, eds. Mechanisms and Consequences of DNA Damage Processing. New York: Alan R. Liss, Inc., pp 413-422.
73. Rinken R and Wackernagel W (1992) Inhibition of the *recBCD*-dependent activation of Chi recombinational hot spots in SOS-induced cells of *Escherichia coli*. J Bacteriol 174:1172-1178.
74. Köppen A, Krobitsch S, Thoms B and Wackernagel W (1995) Interaction with the recombination hot spot χ in vivo converts the RecBCD enzyme of *Escherichia coli* into a χ-independent recombinase by inactivation of the RecD subunit. Proc Natl Acad Sci USA 92:6249-6253.
75. Myers RS, Kuzminov A and Stahl FW (1995) The recombination hotspot χ activates RecBCD recombination by converting *E. coli* to a *recD* mutant phenocopy. Proc Natl Acad Sci USA 92:6244-6248.

76. Brcic-Kostic K, Stojiljkovic I, Salaj-Smic E and Trgovcevic Z (1992) Overproduction of the RecD polypeptide sensitizes *Escherichia coli* cells to γ-radiation. Mutat Res 281:123-127.
77. Dixon DA, Churchill JJ and Kowalczykowski SC (1994) Reversible inactivation of the *Escherichia coli* RecBCD enzyme by the recombination hotspot χ in vitro: Evidence for functional inactivation or loss of the RecD subunit. Proc Natl Acad Sci USA 91:2980-2984.
78. de Vries J and Wackernagel W (1992) Recombination and UV resistance of *Escherichia coli* with the cloned *recA* and *recBCD* genes of *Serratia marcescens* and *Proteus mirabilis*: evidence for an advantage of interspecies combination of *P. mirabilis* RecA protein and RecBCD enzyme. J General Microbiol 138:31-38.
79. Dixon DA and Kowalczykowski SC (1991) Homologous pairing in vitro stimulated by the recombination hotspot, Chi. Cell 66:361-371.
80. De Lucia P and Cairns J (1969) Isolation of an *E. coli* strain with a mutation affecting DNA polymerase. Nature 224:1164-1166.
81. Gross J and Gross M (1969) Genetic analysis of an *E. coli* strain with a mutation affecting DNA polymerase. Nature 224:1166-1168.
82. Kornberg A and Baker TA (1992) DNA Replication. New York: W.H. Freeman and Company, 931 p.
83. Paterson MC, Boyle JM and Setlow RB (1971) Ultraviolet- and X-ray-induced responses of a deoxyribonucleic acid polymerase-deficient mutant of *Escherichia coli*. J Bacteriol 107:61-67.
84. Zieg J, Maples VF and Kushner SR (1978) Recombination levels of *Escherichia coli* K-12 mutants deficient in various replication, recombination, or repair genes. J Bacteriol 134:958-966.
85. Asai T and Kogoma T (1994) D-loops and R-loops: alternative mechanisms for the initiation of chromosome replication in *Escherichia coli*. J Bacteriol 176:1807-1812.
86. Kogoma T, Cadwell GW, Barnard KG and Asai T (1996). Requirement of the DNA replication priming protein, PriA, for homologous recombination and double-strand break repair. J Bacteriol 178: in press.
87. Sandler SJ, Samra HS and Clark AJ (1996). Differential suppression of *priA2::kan* phenotypes in *E. coli* K-12 by mutations in *priA*, *lexA*, and *dnaC*. Genetics: in press.
88. Otsuji N, Iyehara H and Hideshima Y (1974) Isolation and characterization of *Escherichia coli ruv* mutant which forms nonseptate filaments after low doses of ultraviolet light irradiation. J Bacteriol 117:337-344.
89. Lloyd RG, Benson FE and Shurvinton CE (1984) Effect of *ruv* mutations on recombination and DNA repair in *Escherichia coli* K12. Mol Gen Genet 194:303-309.
90. Lloyd RG (1991) Conjugational recombination in resolvase-deficient *ruvC* mutants of *Escherichia coli* K-12 depends on *recG*. J Bacteriol 173: 5414-5418.

91. Lloyd RG and Buckman C (1991) Genetic analysis of the *recG* locus of *Escherichia coli* K-12 and of its role in recombination and DNA repair. J Bacteriol 173:1004-1011.
92. Lloyd RG and Sharples GJ (1991) Molecular organization and nucleotide sequence of the *recG* locus of *Escherichia coli* K-12. J Bacteriol 173:6837-6843.
93. Lloyd RG and Sharples GJ (1993) Processing of recombination intermediates by the RecG and RuvAB proteins of *Escherichia coli.* Nucleic Acids Res 21:1719-1725.
94. Whitby MC, Ryder L and Lloyd RG (1993) Reverse branch migration of Holliday junctions by RecG protein: a new mechanism for resolution of intermediates in recombination and DNA repair. Cell 75:341-350.
95. Sharples GJ, Chan SN, Mahdi AA, Whitby MC and Lloyd RG (1994) Processing of intermediates in recombination and DNA repair: identification of a new endonuclease that specifically cleaves Holliday junctions. EMBO J 13:6133-6142.
96. Mandal TN, Mahdi AA, Sharples GJ and Lloyd RG (1993) Resolution of Holliday intermediates in recombination and DNA repair: indirect suppression of *ruvA*, *ruvB*, and *ruvC* mutations. J Bacteriol 175:4325-4334.
97. Fox MS (1966) On the mechanism of integration of transforming deoxyribonucleate. J Gen Physiol 49:183-196.
98. Whitby MC and Lloyd RG (1995) Branch migration of three-strand recombination intermediates by RecG, a possible pathway for securing exchanges initiated by 3'-tailed duplex DNA. EMBO J 14:3302-3310.
99. Muller B, Tsaneva IR and West SC (1993) Branch migration of Holliday junctions promoted by the *Escherichia coli* RuvA and RuvB proteins. I. Comparison of RuvAB- and RuvB-mediated reactions. J Biol Chem 268:17179-17184.
100. Haynes RH and Kunz BA (1981) DNA repair and mutagenesis in yeast. In: Strathern JN, Jones EW and Broach JR, eds. The Molecular Biology of the Yeast *Saccharomyces.* Cold Spring Harbor, N.Y.: Cold Spring Harbor Laboratory, pp 371-414.
101. Glickman BW (1979) *rorA* mutation of *Escherichia coli* K-12 affects the *recB* subunit of exonuclease V. J Bacteriol 137:658-660.
102. Van Dorp B, Benne R and Palitti F (1975) The ATP-dependent DNAase from *Escherichia coli rorA*: a nuclease with changed enzymatic properties. Biochim Biophys Acta 395:446-454.
103. Glickman BW, Zwenk H, Van Sluis CA and Rorsch A (1971) The isolation and characterization of an X-ray-sensitive ultraviolet-resistant mutant of *Escherichia coli.* Biochim Biophys Acta 254:144-154.
104. Debenham PG and Webb MBT (1988) The isolation and preliminary characterisation of a novel *Escherichia coli* mutant *rorB* with enhanced sensitivity to ionising radiation. Mol Gen Genet 215:161-164.
105. Debenham PG, Webb MBT and Law J (1988) The cloning of the *rorB* gene of *Escherichia coli.* Mol Gen Genet 215:156-160.

106. Daniels DL, Plunkett III G, Burland V and Blattner FR (1992) Analysis of the *Escherichia coli* genome: DNA sequence of the region from 84.5 to 86.5 minutes. Science 257:771-778.
107. Sargentini NJ and Smith KC (1983) Characterization of an *Escherichia coli* mutant (*radB101*) sensitive to γ and UV radiation, and methyl methanesulfonate. Radiat Res 104:109-115.
108. Picksley SM, Attfield PV and Lloyd RG (1984) Repair of DNA double-strand breaks in *Escherichia coli* K12 requires a functional *recN* product. Mol Gen Genet 195:267-274.
109. Sargentini NJ and Smith KC (1986) Role of the *radB* gene in postreplication repair in UV-irradiated *Escherichia coli uvrB*. Mutat Res 166:17-22.
110. Sargentini NJ and Smith KC (1988) Genetic and phenotypic analyses indicating occurrence of the *recN262* and *radB101* mutations at the same locus in *Escherichia coli*. J Bacteriol 170:2392-2394.
111. Finch PW, Chambers P and Emmerson PT (1985) Identification of the *Escherichia coli recN* gene product as a major SOS protein. J Bacteriol 164:653-658.
112. Rostas K, Morton SJ, Picksley SM and Lloyd RG (1987) Nucleotide sequence and LexA regulation of the *Escherichia coli recN* gene. Nucleic Acid Res 15:5041-5049.
113. Blattner FR, Burland V, Plunkett III G, Sofia HJ and Daniels DL (1993) Analysis of the *Escherichia coli* genome. IV. DNA sequence of the region from 89.2 to 92.8 minutes. Nucleic Acids Res 21:5408-5417.
114. Burland V, Plunkett III G, Daniels DL and Blattner FR (1993) DNA sequence and analysis of 136 kilobases of the *Escherichia coli* genome: organizational symmetry around the origin of replication. Genomics 16:551-561.
115. Plunkett III G, Burland V, Daniels DL and Blattner FR (1993) Analysis of the *Escherichia coli* genome. III. DNA sequence of the region from 87.2 to 89.2 minutes. Nucleic Acids Res 21:3391-3398.
116. Sofia HJ, Burland V, Daniels DL, Plunkett III G and Blattner FR (1994) Analysis of the *Escherichia coli* genome. V. DNA sequence of the region from 76.0 to 81.5 minutes. Nucleic Acids Res 22:2576-2586.
117. Burland V, Plunkett III G, Sofia HJ, Daniels DL and Blattner FR (1995) Analysis of the *Escherichia coli* genome VI: DNA sequence of the region from 92.8 through 100 minutes. Nucleic Acids Res 23:2105-2119.
118. Kogoma T and von Meyenburg K (1983) The origin of replication, *oriC*, and the *dnaA* protein are dispensable in stable DNA replication (*sdrA*) mutants of *Escherichia coli* K-12. EMBO J 2:463-468.
119. de Massy B, Fayet O and Kogoma T (1984) Multiple origin usage for DNA replication in *sdrA* (*rnh*) mutants of *Escherichia coli* K-12: initiation in the absence of *oriC*. J Mol Biol 178:227-236.
120. Kobayashi T, Hidaka M and Horiuchi T (1989) Evidence of a *ter* specific binding protein essential for the termination reaction of DNA replication in *Escherichia coli*. EMBO J 8:2435-2441.

121. Bierne H and Michel B (1994) When replication forks stop. Mol Microbiol 13:17-23.
122. Itaya M and Crouch RJ (1991) A combination of RNase H (*rnh*) and *recBCD* or *sbcB* mutations in *Escherichia coli* K12 adversely affects growth. Mol Gen Genet 277:424-432.
123. Lane HED and Denhardt DT (1975) The *rep* mutation. IV. Slower movement of the replication forks in *Escherichia coli rep* strains. J Mol Biol 97:99-112.
124. Uzest M, Ehrlich SD and Michel B (1995) Lethality of *rep recB* and *rep recC* double mutants of *Escherichia coli.* Mol Microbiol 17:1177-1188.
125. Brcic-Kostic K, Salaj-Smic E, Marsic N, Kajic S, Stojilkovic I and Trgovcevic Z (1991) Interaction of RecBCD enzyme with DNA damaged by gamma radiation. Mol Gen Genet 228:136-142.
126. Muller B, Boehmer PE, Emmerson PT and West SC (1991) Action of RecBCD enzyme on Holliday structures made by RecA. J Biol Chem 266:19028-19033.
127. Campbell A, Schneider SJ and Song B (1992) Lambdoid phages as elements of bacterial genomes. Genetica 86:259-267.
128. Joseph JW and Kolodner R (1983) Exonuclease VIII of *Escherichia coli.* II. Mechanism of action. J Biol Chem 258:10418-10424.
129. Hall SD and Kolodner RD (1994) Homologous pairing and strand exchange promoted by the *Escherichia coli* RecT protein. Proc Natl Acad Sci USA 91:3205-3209.
130. Clark AJ, Sharma V, Brenowitz S, Chu CC, Sandler S, Satin L, Templin A, Berger I and Cohen A (1993) Genetic and molecular analyses of the C-terminal region of the *recE* gene from the Rac prophage of *Escherichia coli* K-12 reveal the *recT* gene. J Bacteriol 175:7673-7682.
131. Clark AJ (1971) Toward a metabolic interpretation of genetic recombination of *E. coli* and its phages. Annu Rev Microbiol 25:437-464.
132. Gillen JR, Willis DK and Clark AJ (1981) Genetic analysis of the RecE pathway of genetic recombination in *Escherichia coli* K-12. J Bacteriol 145:521-532.
133. Lovett ST and Clark AJ (1984) Genetic analysis of the *recJ* gene of *Escherichia coli* K-12. J Bacteriol 157:190-196.
134. Kolodner R, Fishel RA and Howard M (1985) Genetic recombination of bacterial plasmid DNA: effect of RecF pathway mutations on plasmid recombination in *Escherichia coli.* J Bacteriol 163:1060-1066.
135. Mahdi AA and Lloyd RG (1989) Identification of the *recR* locus of *Escherichia coli* K-12 and analysis of its role in recombination and DNA repair. Mol Gen Genet 216:503-510.
136. Lloyd RG, Buckman C and Benson FE (1987) Genetic analysis of conjugational recombination in *Escherichia coli* K12 strains deficient in RecBCD enzyme. J Gen Microbiol 133:2531-2538.
137. Lovett ST and Kolodner RD (1989) Identification and purification of a single-stranded DNA-specific exonuclease encoded by the *recJ* gene of *Escherichia coli.* Proc Natl Acad Sci USA 86:2627-2631.

138. Lehman IR and Nussbaum R (1964) The deoxyribonuclease of *Escherichia coli*. V. On the specificity of exonuclease I (phosphodiesterase). J Biol Chem 239:2628-2636.
139. Modrich P (1991) Mechanisms and biological effects of mismatch repair. Annu Rev Genet 25:229-253.
140. Lloyd RG and Buckman C (1985) Identification and genetic analysis of *sbcC* mutations in commonly used *recBC sbcB* strains of *Escherichia coli* K-12. J Bacteriol 164:836-844.
141. Gibson FP, Leach DRF and Lloyd RG (1992) Identification of *sbcD* mutations as cosupressors of *recBC* that allow propagation of DNA palindromes in *Escherichia coli* K-12. J Bacteriol 174:1222-1228.
142. Sharples GJ and Leach DRF (1995) Structural and functional similarities between the SbcCD proteins of *Escherichia coli* and the RAD50 and MRE11 (RAD32) recombination and repair proteins of yeast. Mol Microbiol 17:1215-1220.
143. Naom IS, Morton SJ, Leach DRF and Lloyd RG (1989) Molecular organisation of *sbcC*, a gene that affects genetic recombination and the viability of DNA palindromes in *Escherichia coli* K-12. Nucleic Acids Res 17:8033-8045.
144. Horii ZI and Clark AJ (1973) Genetic analysis of the RecF pathway of genetic recombination in *Escherichia coli* K-12: isolation and characterization of mutants. J Mol Biol 80:327-344.
145. Nakayama H, Nakayama K, Nakayama R, Irino N, Nakayama Y and Hanawalt PC (1984) Isolation and genetic characterization of a thymineless death-resistant mutant of *Escherichia coli* K12: Identification of a new mutation (*recQ1*) that blocks the RecF recombination pathway. Mol Gen Genet 195:474-480.
146. Mendonca VM, Klepin HD and Matson SW (1995) DNA helicases in recombination and repair: construction of a *ΔuvrD ΔhelD ΔrecQ* mutant deficient in recombination and repair. J Bacteriol 177:1326-1335.
147. Matson SW, Bean DW and George JW (1994) DNA helicases: enzymes with essential roles in all aspects of DNA metabolism. BioEssays 16:13-22.
148. Clark AJ and Sandler SJ (1994) Homologous recombination: the pieces begin to fall into place. Crit Rev Microbiol 20:125-142.
149. Dunderdale HJ, Benson FE, Parsons CA, Sharples GJ, Lloyd RG and West SC (1991) Formation and resolution of recombination intermediates by *E. coli* RecA and RuvC proteins. Nature 354:506-510.
150. McClintock B (1932) A correlation of ring-shaped chromosomes with variegation in *Zea mays*. Proc Natl Acad Sci USA 18:677-681.
151. Blakely G, Colloms S, May G, Burke M and Sherratt D (1991) *Escherichia coli* XerC recombinase is required for chromosomal segregation at cell division. New Biol 3:789-798.
152. Clerget M (1991) Site-specific recombination promoted by a short DNA segment of plasmid R1 and by a homologous segment in the terminus region of the *Escherichia coli* chromosome. New Biol 3:780-788.

153. Kuempel PL, Henson JM, Dircks L, Tecklenburg M and Lim DF (1991) *dif*, a *recA*-independent recombination site in the terminus region of the chromosome of *Escherichia coli*. New Biol 3:799-811.
154. Blakely G, May G, McCulloch R, Arciszewska LK, Burke M, Lovett ST and Sherratt DJ (1993) Two related recombinases are required for site-specific recombination at *dif* and *cer* in *E. coli* K12. Cell 75:351-361.
155. Skarstad K and Boye E (1993) Degradation of individual chromosomes in *recA* mutants of *Escherichia coli*. J Bacteriol 175:5505-5509.
156. Leslie NR and Sherratt DJ (1995) Site-specific recombination in the replication terminus region of *Escherichia coli*: functional replacement of *dif*. EMBO J 14:1561-1570.
157. Riley M and Anilionis A (1978) Evolution of the bacterial genome. Ann Rev Microbiol 32:519-560.
158. McKittrick NH and Smith GR (1989) Activation of Chi recombinational hotspots by RecBCD-like enzymes from enteric bacteria. J Mol Biol 210:485-495.
159. Fleischmann RD, Adams MD, White O, et al (1995) Whole-genome random sequencing and assembly of *Haemophilus influenzae* Rd. Science 269:496-512.
160. Sirois S and Szatmari G (1995) Detection of XerC and XerD recombinases in Gram-negative bacteria of the family *Enterobacteriaceae*. J Bacteriol 177:4183-4186.

CHAPTER 6

Single-Strand Annealing: The Alternative Recombinational Repair of Double-Strand Breaks

6.0. SINGLE-STRAND ANNEALING REPAIR OF DOUBLE-STRAND BREAKS

The major recombinational repair pathways in bacterial cells depend on RecA protein. Indeed, the essence of recombinational repair is to synapse a damaged DNA with an intact homologous sequence, and RecA is the only enzyme in bacterial cells that can catalyze such synapsis. However, another way of repairing a damaged DNA is to synapse it with another damaged homologous DNA. Two-strand DNA lesions are always associated with lengthy tracts of single-stranded DNA. Daughter-strand gaps are single-stranded themselves (4.1), double-strand breaks are processed so as to have single-strand overhangs at the double-stranded ends (5.2.1). If these tracts of single-stranded DNA on the homologous molecules are complementary to each other, they can anneal to form a duplex. Annealing of complementary DNA strands does not require participation of RecA since it is promoted by considerable number of other proteins[1] and, in fact, occurs spontaneously. Annealing of complementary single strands associated with nonoverlapping lesions converts a pair of two-strand lesions into a pair of one-strand lesions and serves as a basis for the whole class of simple recombinational repair reactions.

A particular example of such single-strand annealing (SSA) reaction is repair of a double-strand break between two directly repeated sequences, first envisioned by Thomas (1.1.4) (Fig. 6.1). If both ends of the break are processed so that a strand with a particular polarity is degraded while the opposite strand is preserved, the unraveled strands bearing complements of the direct repeat can anneal with each other, reforming duplex DNA. Filling-in the gaps and clipping off single-stranded regions excluded from synapsis completes the double-strand break repair (Fig. 6.1). In contrast to RecA-promoted repair, this recombinational repair cannot be faithful: as a result of it, one copy of

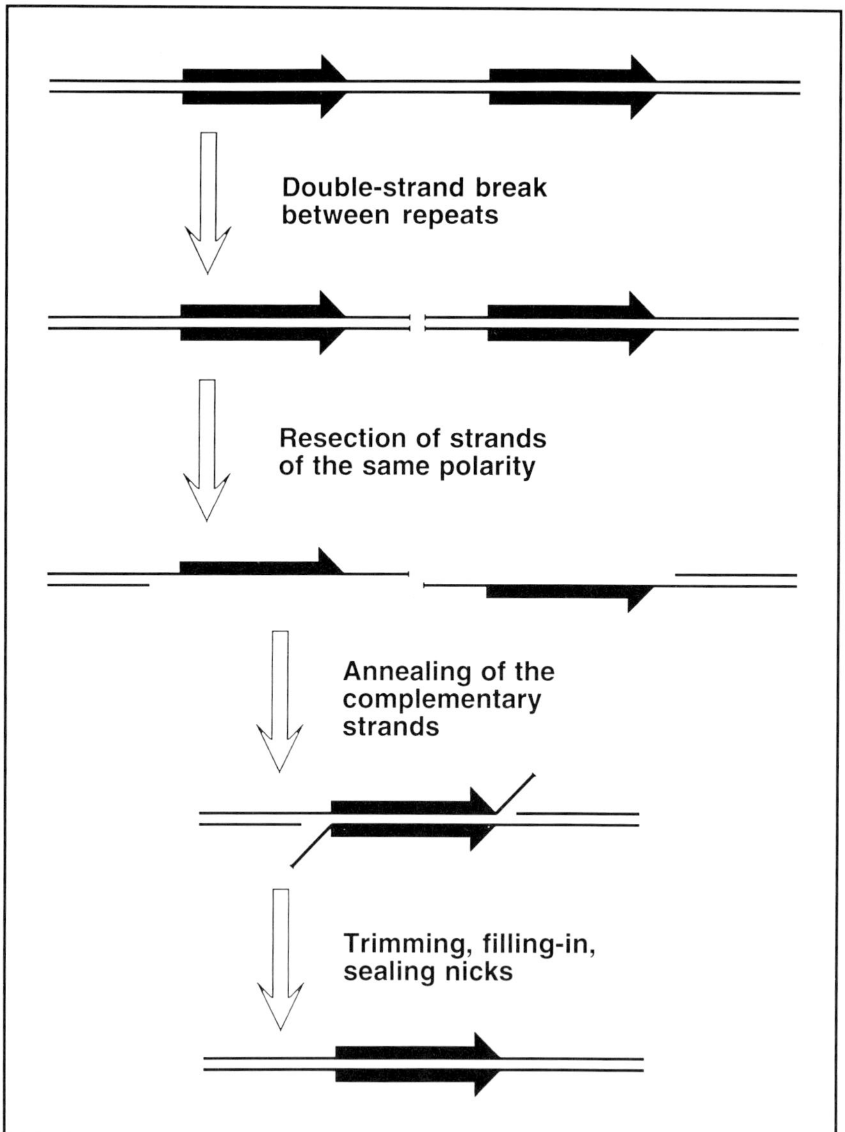

Fig. 6.1. SSA repair of a double-strand break between direct repeats. Direct repeats are shown as black arrows. Explanations in the figure and in the text.

the repeat and the intervening sequence are always lost. Hence, recombination associated with single-strand annealing repair is of the nonconservative type.

6.1. SSA REPAIR – THE BACTERIOPHAGE PANACEA

Because of the intrinsic loss of DNA information, SSA repair can is exercised in bacterial cells only in concatemeric phage and plasmid genomes. Phage lambda recombines well in *recA* mutant cells,[2] although it does not code for a RecA-like synaptase. Lambda, as well as the lambdoid Rac prophage (5.6.1) code for two recombinational repair enzymes; the repair reaction promoted by these enzymes in vitro is of SSA type.[3]

6.1.1. SSA Repair Enzymes of Phage Lambda

To be fully proficient in recombination, phage lambda needs three genes called *exo*, *bet* and *gam*.[4,5] *gam* encodes a small protein with the same name which inhibits all the known activities of RecBCD enzyme.[6,7] In the absence of Gam, linear phage DNA is degraded by the ExoV activity of RecBCD.[8] Analogous proteins are produced by many bacteriophages of *E. coli* replicating their genomes as linear DNA.[9] *exo* codes for an exonuclease which degrades the 5'-ending strand of a linear duplex DNA, producing long 3'-overhangs.[10,11] The enzyme is unable to begin DNA degradation at nicks and is inhibited by completion of strand assimilation[3,12] (Fig. 6.2A). It purifies from infected cells in a complex with another phage protein, called beta.[13]

Beta protein, the product of *bet*, is a single-strand binding protein which promotes annealing of complementary DNA strands.[14] Thus, its sole function could be to anneal complementary strands unraveled by the exonuclease. As mentioned above, annealing of ssDNA occurs spontaneously; nevertheless, Exo^+ $Beta^-$ mutants are deficient in phage recombination,[4] suggesting that Beta might have functions other than just ssDNA annealing. This interpretation is supported by the fact that Beta purifies from infected cells in complex not only with exonuclease, but also with S1 ribosomal protein and with NusA subunit of RNA polymerase.[15,16] S1 is the biggest protein of the small ribosome subunit and is responsible for the mRNA binding.[17] NusA is a regulatory subunit of RNA polymerase — it serves as a transcription elongation factor interacting with other proteins.[18] We entertain the possibility that one of the functions of Beta might be to clear the DNA from transcription-translation complexes ahead of the exonuclease.

6.1.2. SSA Repair Enzymes of Rac Prophage

Rac is a cryptic lambdoid prophage residing in the *E. coli* chromosome (5.6.1). It codes for at least two SSA repair genes which are usually silent, but can be activated by *sbcA* mutations. These two recombinational genes can also be picked up by a lambda deleted for its

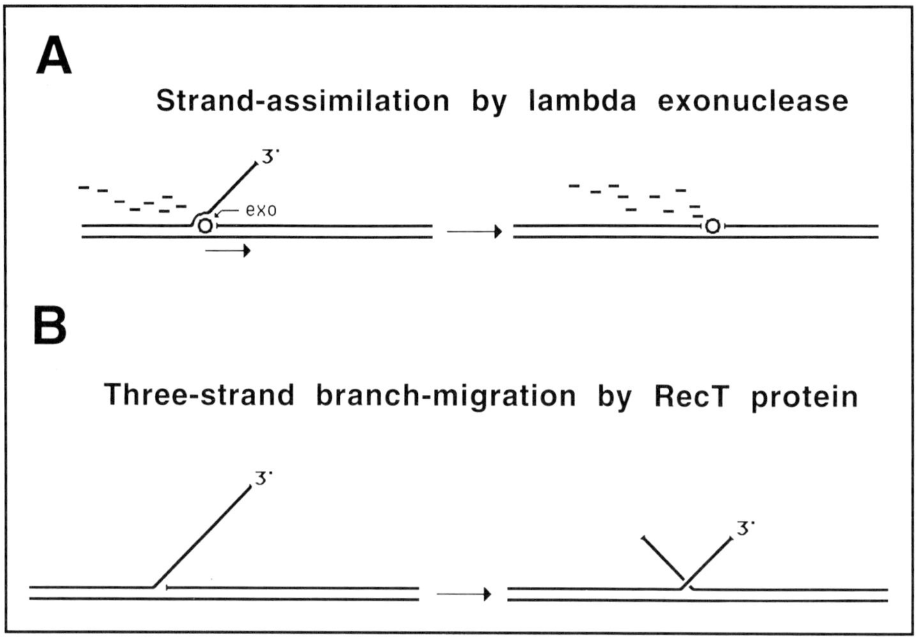

Fig. 6.2. Some reactions promoted by SSA repair proteins. (A) Strand-assimilation catalyzed by lambda exonuclease. Lambda exonuclease is marked by "exo"; small arrow below the DNA shows the direction of exonucleolytic degradation. The exonuclease will degrade the 5'-ending strand of the duplex until the branching strand is completely assimilated into the duplex.[12] Exonuclease stops its progress when strand assimilation generates a single-strand interruption, accounting for the inability of lambda exonuclease to start DNA degradation from single-strand interruptions. (B) Three-stranded branch migration catalyzed by RecT protein. If RecT-promoted annealing of complementary strands is blocked by a duplex region in one of the molecules, the protein is able to catalyze strand exchange, displacing the resident strand with the incoming strand. The reaction resembles branch migration promoted by RecA protein. However, in contrast to RecA-promoted strand exchange, RecT is able to start this reaction only if participating molecules have single-stranded complementary regions.[24]

own recombinational genes, producing a recombination-proficient revertant.[5,19,20] The product of *recE* gene is exonuclease VIII, which degrades the 5'-ending strand of a linear duplex DNA and, hence, is analogous to lambda exonuclease.[21] However, two genes do not share significant homology, and two enzymes are very different in everything including molecular weight: while lambda exonuclease is a 26 kDa dwarf, ExoVIII is a 140 kDa monster.[22]

Beta protein of phage lambda also has its counterpart in Rac prophage, called RecT protein. RecT protein promotes annealing of complementary single strands and also catalyzes a three-strand branch migration when annealing encounters a double-stranded region[23,24] (Fig. 6.2B). The biological significance of the latter reaction is not obvious, although it is basically similar to the strand assimilation reaction catalyzed by lambda exonuclease (Fig. 6.2A).

6.1.3. What SSA Repair Can Do and What It Cannot Do

In which circumstances can SSA repair substitute for double-strand end repair? Generalized transduction is a gene conversion of the host allele with a chromosomal piece of a different genotype, brought in by a bacteriophage (2.1.2). Mechanistically, it requires "double-strand end repair" of the linear transducing DNA with a circular chromosome. Consistent with the notion that any strand invasion into a circular duplex requires RecA, the frequency of generalized transduction is reduced two to three orders of magnitude in *recA* mutants.[25,26] Remarkably, the Red recombination system of phage lambda is able to catalyze a low level of transduction in the absence of RecA.[27,28] We suggest that this Red-promoted recombination occurs at replication forks, where single-stranded pieces of DNA could be gradually assimilated via annealing with the template for the lagging strand DNA synthesis (Fig. 6.3). The scheme predicts that the RecA-independent Red-mediated transduction will be influenced by position of replication forks and will be inhibited if DNA replication is temporarily blocked.

The low level of RecA-independent transduction is the only repair transaction the Red or the RecE pathways can do in the chromosome in the absence of RecA. SSA repair can not mend the chromosomal DNA lesions in *recA* mutants,[29] underscoring its usefulness only for repair of double-strand breaks in highly repetitive genomes. What is the role of SSA repair in the phage life cycle?

Early in infection phage lambda replicates as a big plasmid, producing circular monomeric copies of its genome. Later phage replication switches from theta- to sigma-mode, generating linear several-phage-genomes-long arrays, which are subject to packaging. The packaging of lambda genome starts and ends at a site on phage chromosome called *cos*. The sequences important for the *cos*-recognition by packaging machinery are mapped to both sides of the actual double-strand break site.[30] Therefore, only genomes bracketed by two intact *cos* sites can be packaged in vivo. Packaging of lambda is processive,[31] so longer arrays yield more packageable genomes. SSA repair probably acts on the products of sigma-replication and on leftovers from packaging to increase their length and, therefore, the yield of phage genomes (Fig. 6.4). This notion is consistent with the fact that *red* mutants of lambda produce only 20-30% as many progeny as WT phages.[5,8]

6.2. EXPERIMENTAL SET-UPS IN WHICH SSA REPAIR LOOKS LIKE DOUBLE-STRAND END REPAIR

Due to the high efficiency stemming from its simplicity, single-strand annealing repair is easy to demonstrate in vivo with model systems. Besides, the yield of SSA repair is higher with smaller DNA molecules, while the reverse seems to be true for the double-strand end repair. Finally, in certain experimental set-ups, recombinational products do not allow one to distinguish between these two models

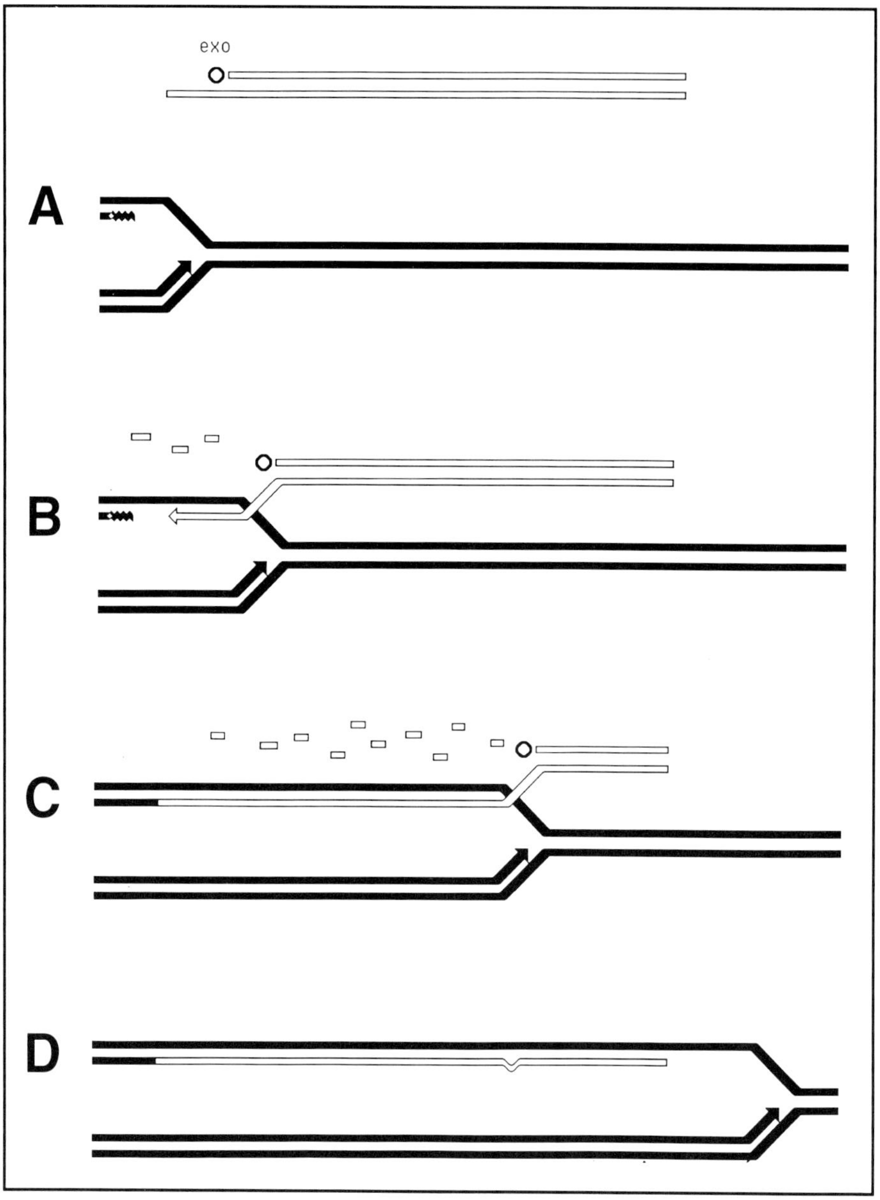

Fig. 6.3. A mechanism for Red-mediated transduction in the absence of RecA. Open lines – the incoming duplex DNA, solid lines – the host chromosome, arrows indicate 3'-ends, circle – lambda exonuclease. (A) Transducing DNA is degraded by lambda exonuclease, while a replication fork is passing through the homologous region on the chromosome. (B) Beta protein catalyzes annealing of the incoming strand with the template for the lagging strand DNA synthesis. (C) The assimilation of the incoming strand into the host DNA catalyzed by combined action of lambda exonuclease and beta protein. (D) The strand is completely assimilated. There is a heterology (shown as a bump) between the transducing DNA strand and the host strand. This heterology is, probably, resolved by the next round of DNA replication since the incoming DNA is likely to be fully methylated (a poor substrate for the methyl-directed mismatch repair (1.1.1)).

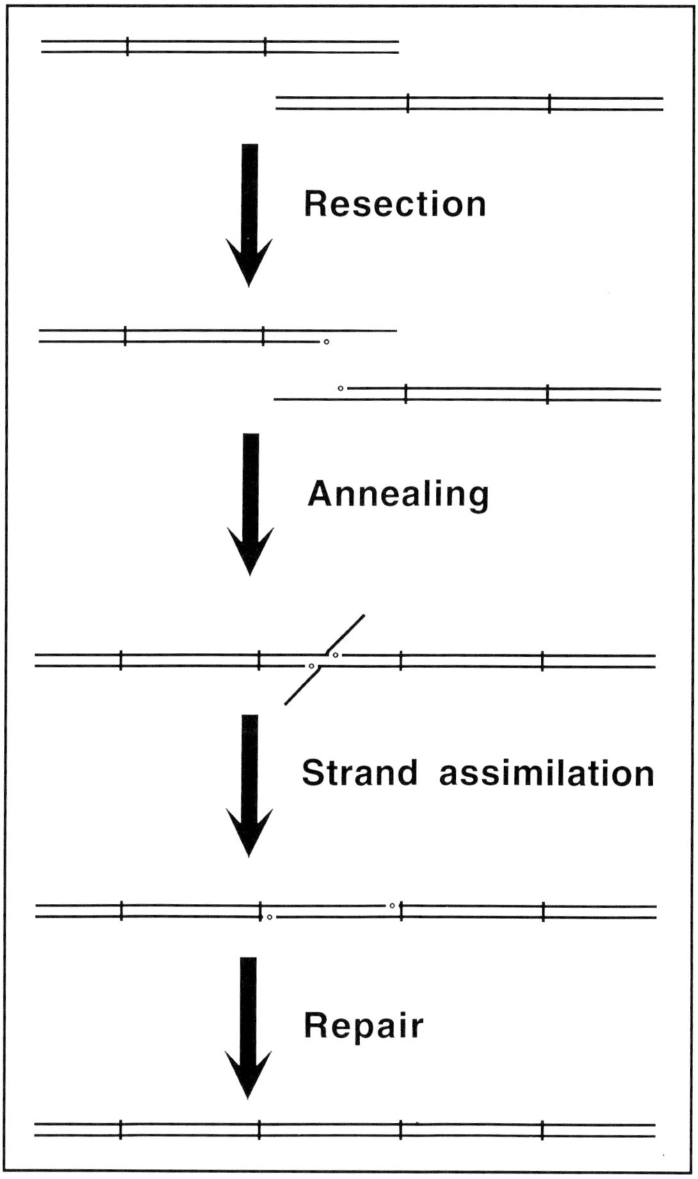

Fig. 6.4. A possible role of SSA repair in increasing packageability of lambda DNA. Products of late DNA replication – linear concatemeric pieces – are represented by the double lines; genome equivalents in them are marked by the short vertical lines. A packageable genome is bracketed by two intact vertical lines. Thus, from each initial DNA piece, only one phage genome can be packaged (total two), while the product of SSA repair contains three packageable genome equivalents.

of double-strand break repair. All this led to the appearance of several reports, whose authors have been interpreting their findings in terms of double-strand end repair, although their results were more compatible with single-strand annealing repair. Below we will discuss these cases, challenging the authors' conclusions.

6.2.1. Phage Lambda Crosses in the Absence of DNA Replication

Phage lambda is unable to package intact genomes out of monomeric chromosomes in vivo[32] (6.1.3) and switches its DNA replication to the rolling-circle mode late in infection, generating concatemeric arrays of the phage genomes as substrates for packaging.[8] When lambda's replication is tightly blocked, phage has to recombine to generate packageable chromosomes. If the phage is *red gam* (6.1.1) and harbors Chi (5.3.4) it still shows a decent yield from a replication-blocked infection, almost all the phages coming out of the mixed infection being of recombinant nature.[33,34] This is now understood in terms of efficient RecA- and RecBCD-promoted invasion of phage chromosomes linearized at *cos* into the remaining circular phage genomes.[35] The same could be true for the WT lambda infection in a RecA$^+$ host.[36] In contrast, when replication of WT phage is blocked in a *recA* host, very few progeny are produced, indicating that in this set-up, lambda's own recombination system, Red, cannot generate packageable chromosomes (Fig. 3A in ref. 32; Fig. 5 in ref. 37). Apparently, lambda Red cannot recombine chromosomes linearized at *cos* with circular chromosomes, — that is, lambda Red does not promote double-strand end repair. Later, certain Red$^+$ RecA$^-$ crosses were found to be proficient in recombination even in the absence of DNA replication, allowing the authors to explain their results in terms of double-strand end repair, substituting Beta protein for RecA. Interestingly, in all these cases, the second double-strand break nonallelic to the regular *cos* site has been provided, either as a centrally-placed *cos* site, or as a restriction cut.[36,38,39] Thus, phages could recombine by single-strand annealing as shown in Fig. 6.1 and Fig. 6.4. This alternative interpretation is consistent with several features of these crosses. First, the recombination was eliminated in the absence of the second, nonallelic double-strand break.[36,38] Second, almost all recombinant phages carried the information from both parents (were heteroduplex) across all the region between the two double-strand breaks;[36] such long strand-exchange products cannot be expected for double-strand end repair. Third, the recombination was independent of RecA protein. None of these observations could be adequately explained by the double-strand end repair model.

6.2.2. Double-Strand Break Repair in Plasmids with Direct Repeats

Formation of deletions between direct repeats in *E. coli* plasmids does not require RecA protein in *recBC sbcA* genetic background (5.6.1), where the recombinational system of Rac prophage is expressed.[40,41] In this background, plasmids are thought to form linear multimers similar to DNA of phage lambda[42] and are able to recircularize in a RecA-independent manner with the formation of a deletion between repeats[43] (Fig. 6.5). Indeed, if transformed with a linear dimer plasmid, a *recBC*

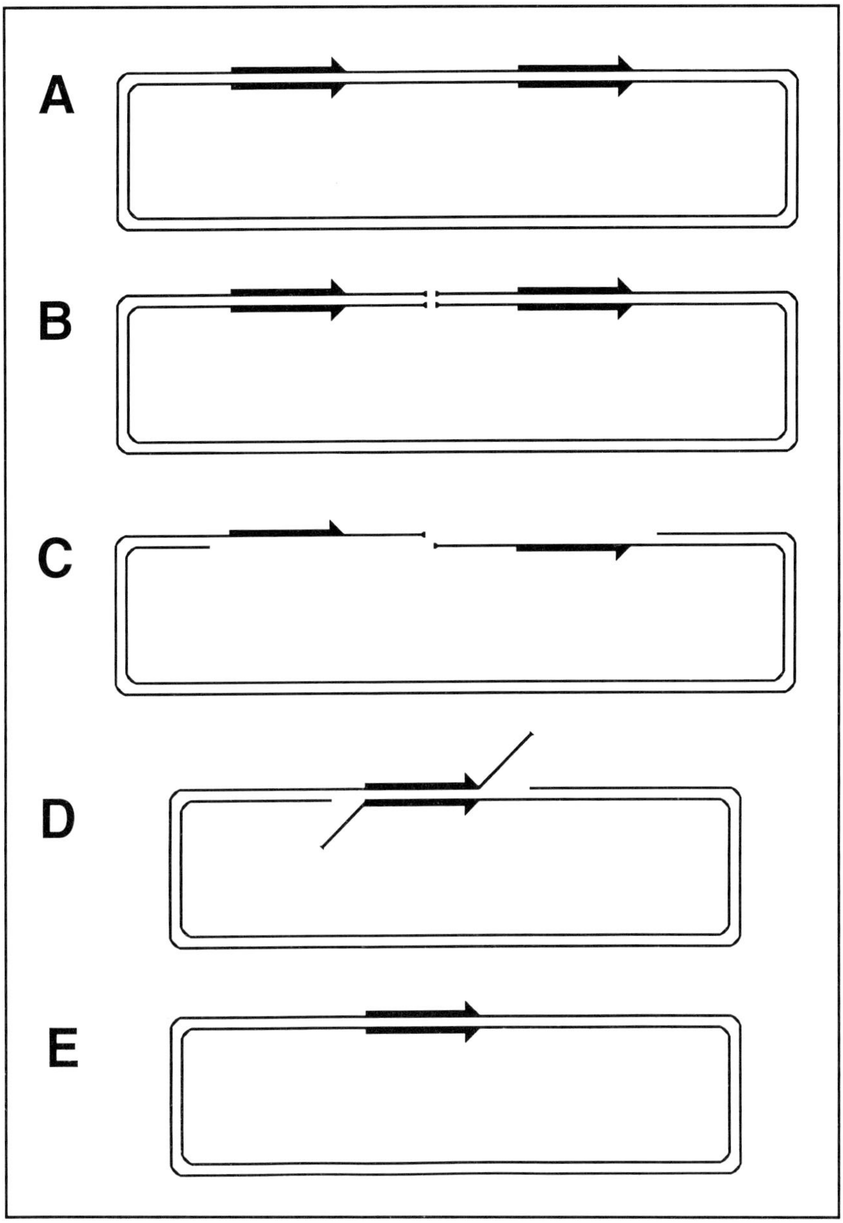

Fig. 6.5. SSA repair of double-strand breaks between direct repeats in plasmid DNA. Basically, this figure is a repetition of Fig. 6.1, but applied to a circular genome. Direct repeats are shown by arrows. (A) A plasmid with direct repeats. (B) A double-strand break is introduced between repeats. (C) Strand-specific resection of double-strand ends. (D) Annealing of the unraveled complements of the direct repeat. (E) Removing the DNA excluded from the synapsis, sealing the interruptions. The resulting recircularized plasmid suffered a deletion.

sbcA strain recircularizes the plasmid with efficiency approaching unity.[44] This repair is dependent on RecE nuclease and independent of any other tested Rec proteins (RecT had not been tested), including RecA.[44,45] Long heterologies at one of the ends of such linear plasmids do not preclude efficient recircularization.[46,47] The last finding was interpreted to signify a mechanism based on double-strand end invasion, where the role of synaptase is played by RecT protein.[46,47] However, since RecT protein alone does not promote RecA-like synapsis,[24] SSA repair remains the most economical way to explain these data. SSA repair should be able to tolerate substantial heterologies at the ends, if the single-strand "whiskers" are removed by ssDNA-specific nucleases (Fig. 6.1).

6.2.3. Double-Strand Break Repair in Plasmids with Inverted Repeats

In a plasmid with long inverted repeats each containing a drug-resistance gene with nonallelic mutations, recombination can restore the functional drug-resistance gene in one of the repeats.[48] When WT or *recBC sbcA* cells are transformed with such a plasmid and immediately plated for the drug-resistance, recombinants are infrequent.[49,50] However, if *recBC sbcA* cells are transformed with the plasmid that was cut in the middle of one of the repeats, the recombination frequency goes up 100-fold (WT cells show no increase).[49-51] Two recombinational products are formed: 30% of recombinants have the original configuration of markers, while the other 70% have the markers in the flipped orientation (cross-over) (Fig. 6.6).[49] This double-strand break-stimulated recombination absolutely requires Rac prophage-encoded RecE and RecT proteins and is independent of any host recombinational gene, including *recA*.[50,51] The ratio of the two recombinational products is more-or-less constant across the "recombinational alphabet" with the exception of *recJ* mutants, in which the ratio of crossover to noncrossover products is reversed.[51] Cloned recombinational genes of phage lambda also promote this double-strand break repair reaction.[52] RecA-independence notwithstanding, authors explain their results in terms of conservative double-strand break repair, assigning the role of synaptase to RecT protein of Rac or to Beta protein of lambda.[49-52] However, formation of the two types of recombinants in their system can be explained by single-strand annealing repair as well (Fig. 6.7); this explanation is more attractive because it does not require participation of RecA-like activity. SSA repair also explains the influence of RecJ exonuclease, which could degrade the 5' end of the intermediate in Fig. 6.7F, preventing the formation of the noncrossover product. Hence, in the absence of RecJ nuclease, there is no discrimination against the noncrossover products, resulting in the flipped ratio of crossovers and noncrossovers. A similar experimental set-up has been recently used to study double-strand break repair in yeast

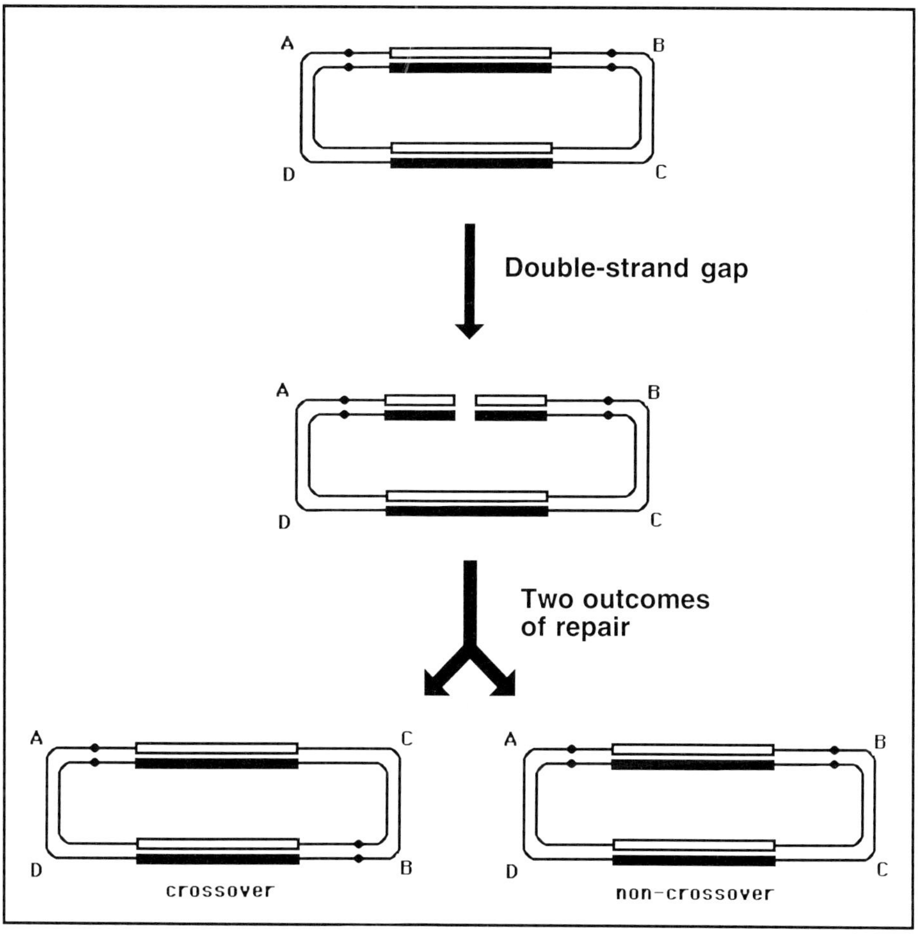

Fig. 6.6. Two products of double-strand break repair in a plasmid with inverted repeats. Inverted repeats are shown as black-and-white duplexes on the opposite sides of the plasmid circle. One side of the circle is marked by knots on DNA strands to facilitate detection of inversion. Letters A, B, C and D serve the same purpose. Explanations in the figure.

with the expected conclusion that DSB repair in plasmids does not require yeast RecA-like functions.[53] Again, the simplest interpretation of this result is that it is SSA repair, as in Fig. 6.7.

SUMMARY

Single-strand annealing repair of double-strand breaks is a simple and extremely efficient reaction, which is possible whenever a double-strand break is flanked by directly repeated sequences. However, the efficiency comes at a cost of a loss of the DNA sequence between repeats together with one of the repeats. Since bacterial genomes are composed mostly of unique sequences, bacteria can not afford SSA

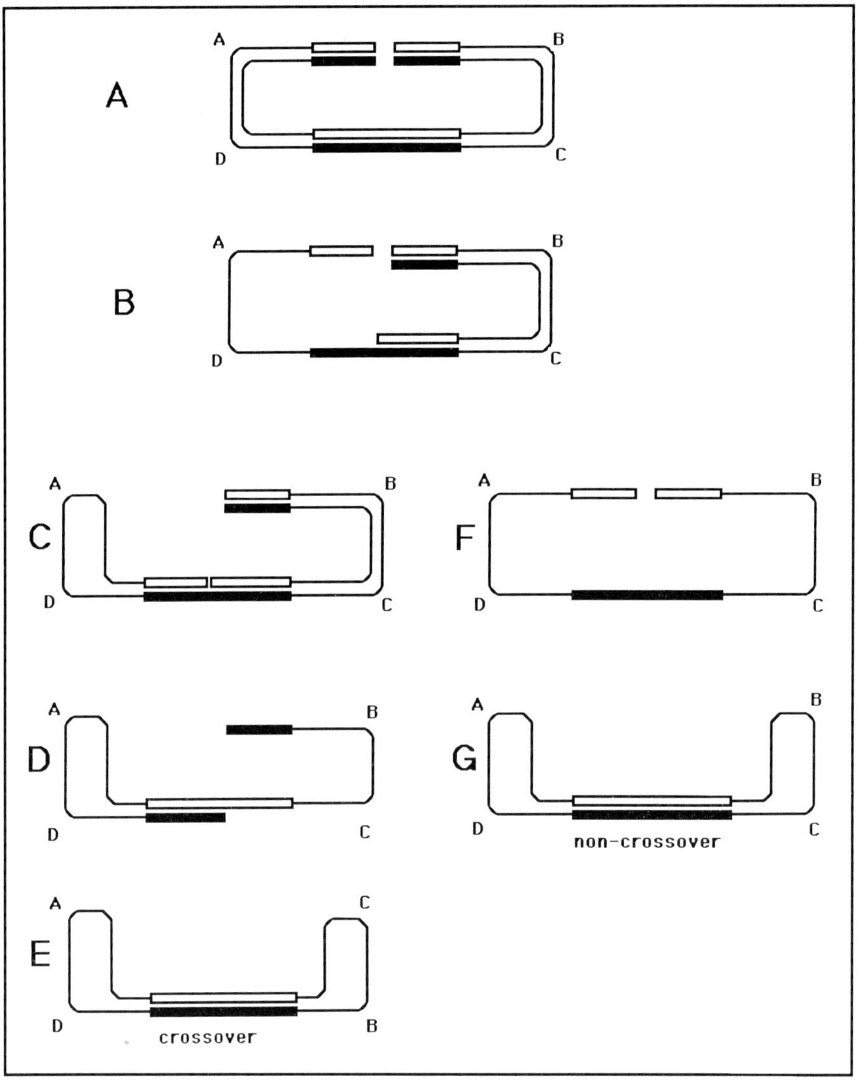

Fig. 6.7. SSA repair of a double-strand break in a plasmid with inverted repeats. Inverted repeats are shown as black-and-white duplexes on the opposite sides of the plasmid circle. Letters A, B, C and D serve to facilitate detection of inversion. (A) A plasmid carrying two inverted repeats with a double-strand break in the middle of one of the repeats. (B) The degradation of 5'-ending strand from one of the ends up to the middle of the intact inverted repeat. (C) The unraveled strand anneals on itself at the inverted repeat. The degradation stops, and the nick is sealed, resulting in a linear molecule with a loop on one of the ends. (D) Degradation of the same polarity from the other end until the middle of the intact repeat is reached. (E) The unraveled strand again anneals on itself using the inverted repeat. The resulting single-stranded circular DNA is then converted to a double-stranded form (not shown). It has markers on the opposite sides of the inverted repeats flipped, as if there was a conservative repair with associated crossing-over (Fig. 6.6). (F) Alternatively, the degradation in "B" goes all the way to the other end of the molecule, rendering it completely single-stranded. (G) The broken repeat anneals with the intact repeat and is repaired off it. After the degraded strand is resynthesized (not shown), this will be a product resembling a conservative double-strand break repair without crossing-over (Fig. 6.6).

QUESTIONNAIRE

Receive a FREE BOOK of your choice

Please help us out—Just answer the questions below, then select the book of your choice from the list on the back and return this card.

R.G. Landes Company publishes five book series: *Medical Intelligence Unit, Molecular Biology Intelligence Unit, Neuroscience Intelligence Unit, Tissue Engineering Intelligence Unit* and *Biotechnology Intelligence Unit.* We also publish comprehensive, shorter than book-length reports on well-circumscribed topics in molecular biology and medicine. The authors of our books and reports are acknowledged leaders in their fields and the topics are unique. Almost without exception, there are no other comprehensive publications on these topics.

Our goal is to publish material in important and rapidly changing areas of bioscience for sophisticated scientists. To achieve this goal, we have accelerated our publishing program to conform to the fast pace in which information grows in bioscience. Most of our books and reports are published within 90 to 120 days of receipt of the manuscript.

Please circle your response to the questions below.

1. We would like to sell our *books* to scientists and students at a deep discount. But we can only do this as part of a prepaid subscription program. The retail price range for our books is \$59-\$99. Would you pay \$196 to select four *books* per year from any of our Intelligence Units–\$49 per book–as part of a prepaid program?

   Yes **No**

2. We would like to sell our *reports* to scientists and students at a deep discount. But we can only do this as part of a prepaid subscription program. The retail price range for our reports is \$39-\$59. Would you pay \$145 to select five *reports* per year–\$29 per report–as part of a prepaid program?

   Yes **No**

3. Would you pay \$39–the retail price range of our books is \$59-\$99–to receive any single book in our Intelligence Units if it is spiral bound, but in every other way identical to the more expensive hardcover version?

   Yes **No**

To receive your free book, please fill out the shipping information below, select your free book choice from the list on the back of this survey and mail this card to:

R.G. Landes Company, 909 S. Pine Street, Georgetown, Texas 78626 U.S.A.

Your Name ______________________________

Address ______________________________

City ____________________ State/Province: ____________________

Country: ____________________ Postal Code: ____________________

My computer type is Macintosh______ ; IBM-compatible ______ ; Other ______

Do you own ____ or plan to purchase ___ a CD-ROM drive?

repair in their chromosomes. However, this repair strategy is widely used by bacteriophages, whose linear concatemeric chromosomes are perfect substrates for SSA repair.

The simple enzymatic requirements of SSA repair are best illustrated with phage lambda and Rac prophage recombinational systems. In both cases, only two enzymes are required. One is a duplex DNA-specific exonuclease that selectively degrades strands ending 5', generating 3'-overhangs. The other is an enzyme catalyzing annealing of complementary DNA strands. The final processing of recombination intermediates is likely to require sealing the single-strand interruptions by DNA ligase[3] which is provided by the host. The role of SSA repair in phage life cycle could be recycling of unpackageable DNA.

SSA repair is easy to model in vivo with plasmids carrying repeats. Frequently, the products of these reactions are indistinguishable from presumable products of RecA-promoted conservative double-strand break repair. Sometimes RecA-independence of these in vivo reactions is the only warning that the process in question is SSA repair rather than double-strand end repair. In these cases, it is important to follow Occam's razor rule and choose the simplest explanation out of two possibilities.

SSA repair concludes our tour of recombinational repair reactions and their alternatives in *E. coli*. The next and final chapter of the book considers applicability of the recombinational repair pattern in *E. coli* to recombinational repair in eukaryotes.

References

1. Kowalczykowski SC and Eggleston AK (1994) Homologous pairing and DNA strand-exchange proteins. Annu Rev Biochem 63:991-1043.
2. Brooks K and Clark AJ (1967) Behavior of λ bacteriophage in a recombination deficient strain of *Escherichia coli*. J Virol 1:283-293.
3. Cassuto E, Lash T, Sriprakash KC and Radding CM (1971) Role of exonuclease and β protein of phage λ in genetic recombination. V. Recombination of λ DNA in vitro. Proc Natl Acad Sci USA 68:1639-1643.
4. Shulman MJ, Hallick LM, Echols H and Signer ER (1970) Properties of recombination-deficient mutants of bacteriophage lambda. J Mol Biol 52:501-520.
5. Zissler J, Signer E and Schaefer F (1971) The role of recombination in growth of bacteriophage lambda. I. The gamma gene. In: Hershey AD, ed. The Bacteriophage Lambda. Cold Spring Harbor: Cold Spring Harbor Laboratory Press, pp 455-468.
6. Unger RC and Clark AJ (1972) Interaction of the recombination pathways of bacteriophage λ and its host *Escherichia coli* K12: effects on exonuclease V activity. J Mol Biol 70:539-548.
7. Murphy KC (1991) λ Gam protein inhibits the helicase and χ-stimulated recombination activities of *Escherichia coli* RecBCD enzyme. J Bacteriol 173:5808-5821.

8. Enquist LW and Skalka A (1973) Replication of bacteriophage λ DNA dependent on the function of host and viral genes. I. Interaction of *red, gam,* and *rec.* J Mol Biol 75:185-212.
9. Sakaki Y (1974) Inactivation of the ATP-dependent DNase of *Escherichia coli* after infection with double-stranded DNA phages. J Virol 14: 1611-1612.
10. Little JW (1967) An exonuclease induced by bacteriophage λ. II. Nature of the enzymatic reaction. J Biol Chem 242:679-686.
11. Carter DM and Radding CM (1971) The role of exonuclease and β protein of phage λ in genetic recombination. II. Substrate specificity and the mode of action of λ exonuclease. J Biol Chem 246:2502-2510.
12. Cassuto E and Radding CM (1971) Mechanism for the action of λ exonuclease in genetic recombination. Nature New Biol 229:13-16.
13. Radding CM, Rosenzweig J, Richards F and Cassuto E (1971) Separation and characterization of exonuclease, β protein, and a complex of both. J Biol Chem 246:2510-2512.
14. Muniyappa K and Radding CM (1986) The homologous recombination system of phage λ: pairing activities of β protein. J Biol Chem 261:7472-7478.
15. Muniyappa K and Mythili E (1993) Phage lambda beta protein, a component of general recombination, is associated with host ribosomal S1 protein. Biochem Mol Biol Internat 31:1-11.
16. Venkatesh TV and Radding CM (1993) Ribosomal protein S1 and NusA protein complexed to recombination protein β of phage λ. J Bacteriol 175:1844-1846.
17. Tzareva NV, Makhno VI and Boni IV (1994) Ribosome-messenger recognition in the absence of the Shine-Dalgarno interactions. FEBS Lett 337:189-194.
18. Das A (1993) Control of transcription termination by DNA-binding proteins. Annu Rev Biochem 62:893-930.
19. Gottesman MM, Gottesman ME, Gottesman S and Gellert M (1974) Characterization of bacteriophage λ reverse as an *Escherichia coli* phage carrying a unique set of host-derived recombination functions. J Mol Biol 88:471-487.
20. Kaiser K and Murray NE (1979) Physical characterization of the "Rac prophage" in *E. coli* K-12. Mol Gen Genet 175:159-174.
21. Joseph JW and Kolodner R (1983) Exonuclease VIII of *Escherichia coli.* II. Mechanism of action. J Biol Chem 258:10418-10424.
22. Joseph JW and Kolodner R (1983) Exonuclease VIII of *Escherichia coli.* I. Purification and physical properties. J Biol Chem 258:10411-10417.
23. Hall SD, Kane MF and Kolodner RD (1993) Identification and characterization of the *Escherichia coli* RecT protein, a protein encoded by the *recE* region that promotes renaturation of homologous single-stranded DNA. J Bacteriol 175:277-287.
24. Hall SD and Kolodner RD (1994) Homologous pairing and strand exchange promoted by the *Escherichia coli* RecT protein. Proc Natl Acad Sci USA 91:3205-3209.

25. Hertman I and Luria SE (1967) Transduction studies on the role of a *rec*+ gene in the ultraviolet induction of prophage lambda. J Mol Biol 23:117-133.
26. Zieg J and Kushner SR (1977) Analysis of genetic recombination between two partially deleted lactose operons of *Escherichia coli* K12. J Bacteriol 131:123-132.
27. Wackernagel W and Radding CM (1974) Transformation and transduction of *Escherichia coli*: the nature of recombinants formed by Rec, RecF, and λ Red. In: Grell RF, ed. Mechanisms in Recombination. New York: Plenum Press, pp 111-122.
28. Weisberg RA and Sternberg N (1974) Transduction of *recB*− host is promoted by λ *Red*+ function. In: Grell RF, ed. Mechanisms in Recombination. New York: Plenum Press, pp 107-109.
29. Gillen JR, Willis DK and Clark AJ (1981) Genetic analysis of the RecE pathway of genetic recombination in *Escherichia coli* K-12. J Bacteriol 145:521-532.
30. Becker A and Murialdo H (1990) Bacteriophage λ DNA: the beginning of the end. J Bacteriol 172:2819-2824.
31. Feiss M, Sippy J and Miller G (1985) Processive action of terminase during sequential packaging of bacteriophage λ chromosome. J Mol Biol 186:759-771.
32. Stahl FW, McMilin KD, Stahl MM, Malone RE, Nozu Y and Russo VEA (1972) A role for recombination in the production of "free-loader" lambda bacteriophage particles. J Mol Biol 68:57-67.
33. Lam ST, Stahl MM, McMilin KD and Stahl FW (1974) Rec-mediated recombinational hot spot activity in bacteriophage lambda. II. A mutation which causes hot spot activity. Genetics 77:425-433.
34. McMilin KD, Stahl MM and Stahl FW (1974) Rec-mediated recombinational hot spot activity in bacteriophage lambda. I. Hot spot activity associated with spi− deletions and *bio* substitutions. Genetics 77:409-423.
35. Stahl FW, Thomason LC, Siddiqi I and Stahl MM (1990) Further test of a recombination model in which χ removes the RecD subunit from the RecBCD enzyme of *Escherichia coli*. Genetics 126:519-533.
36. Stahl FW, Fox MS, Faulds D and Stahl MM (1990) Break-join recombination in phage λ. Genetics 125:463-474.
37. Stahl FW, McMilin KD, Stahl MM and Nozu Y (1972) An enhancing role for DNA synthesis in formation of bacteriophage lambda recombinants. Proc Natl Acad Sci USA 69:3598-3601.
38. Stahl FW, Kobayashi I and Stahl MM (1985) In phage λ, *cos* is a recombinator in the Red pathway. J Mol Biol 181:199-209.
39. Stahl FW and Stahl MM (1985) Non-reciprocal crossing over in pahge λ. J Genet 64:31-39.
40. Fishel RA, James AA and Kolodner R (1981) RecA-independent general recombination of plasmids. Nature 294:184-186.
41. Laban A and Cohen A (1981) Interplasmidic and intraplasmidic recombination in *Escherichia coli* K-12. Mol Gen Genet 184:200-207.

42. Cohen A and Clark AJ (1986) Synthesis of linear plasmid multimers in *Escherichia coli* K-12. J Bacteriol 167:327-335.
43. Silberstein Z, Shalit M and Cohen A (1993) Heteroduplex strand-specificity in restriction-stimulated recombination by the RecE pathway of *Escherichia coli.* Genetics 133:439-448.
44. Symington LS, Morrison P and Kolodner R (1985) Intramolecular recombination of linear DNA catalyzed by the *Escherichia coli* RecE recombination system. J Mol Biol 186:515-525.
45. Luisi-DeLuca C, Lovett ST and Kolodner RD (1989) Genetic and physical analysis of plasmid recombination in *recB recC sbcB* and *recB recC sbcA Escherichia coli* K-12 mutants. Genetics 122:269-278.
46. Luisi-DeLuca C and Kolodner RD (1992) Effect of terminal non-homology on intramolecular recombination of linear plasmid substrates in *Escherichia coli.* J Mol Biol 227:72-80.
47. Silberstein Z, Tzfati Y and Cohen A (1995) Primary products of break-induced recombination by *Escherichia coli* RecE pathway. J Bacteriol 177:1692-1698.
48. Yamamoto K, Yoshikura H, Takahashi N and Kobayashi I (1988) Apparent gene conversion in an *Escherichia coli* rec^+ strain is explained by multiple rounds of reciprocal crossing-over. Mol Gen Genet 212:393-404.
49. Kobayashi I and Takahashi N (1988) Double-stranded gap repair of DNA by gene conversion in *Escherichia coli.* Genetics 119:751-757.
50. Takahashi NK, Kusano K, Yokochi T, Kitamura Y, Yoshikura H and Kobayashi I (1993) Genetic analysis of double-strand break repair in *Escherichia coli.* J Bacteriol 175:5176-5185.
51. Kusano K, Sunohara Y, Takahashi N, Yoshikura H and Kobayashi I (1994) DNA double-strand break repair: genetic determinants of flanking crossing-over. Proc Natl Acad Sci USA 91:1173-1177.
52. Takahashi N and Kobayashi I (1990) Evidence for the double-strand break repair model of bacteriophage λ recombination. Proc Natl Acad Sci USA 87:2790-2794.
53. Sugawara N, Ivanov EL, Fishman-Lobell J, Ray BL, Wu X and Haber JE (1995) DNA structure-dependent requirements for yeast *RAD* genes in gene conversion. Nature 373:84-86.

CHAPTER 7

Recombinational Repair in Eukaryotes

The preceding chapters unraveled our current understanding of the *E. coli*'s recombinational repair and back-up pathways. Two-strand lesions in the chromosomal DNA of *E. coli* are repaired via two RecA-dependent pathways (Table 7.1). Daughter-strand gaps are repaired by the RecF pathway, which in WT cells operates on DNA molecules lacking free double-strand ends. If the RecA-dependent daughter-strand gap repair is impossible because of the absence or inaccessibility of an intact homologous sequence, translesion DNA synthesis regulated by RecA protein is used as a back-up. Double-stranded ends are converted into replication forks by the RecBCD pathway, which is therefore thought to mend disintegrated replication forks as well as double-strand breaks. If the RecA-dependent double-strand end repair is impossible, the entire linear tail is likely to be eliminated via the RecBCD-promoted degradation. When RecBCD is inactivated to be replaced by phage recombinases, double-strand breaks in repetitive genomes can be mended independently of RecA by single-strand annealing repair. In response to a massive DNA damage, *E. coli* cells increase their capacity for damage repair and tolerance, but do not actively inhibit DNA replication. Recombinational repair and damage responses in other eubacteria are likely to be variations of the same general theme. To what extent is the bacterial paradigm applicable to repair of two-strand DNA damage in eukaryotes?

7.0. GENERAL RESPONSE OF EUKARYOTIC CELLS TO DNA DAMAGE

In contrast to *E. coli*, eukaryotic cells respond to DNA damage by blocking cell-cycle progress. If cells are in G_1 phase, DNA damage delays the following S-phase; if cells are in S-phase, DNA damage inhibits DNA synthesis; if cells are in G_2 phase, DNA damage delays mitosis. This inhibition is effected on the global as well as on the

***Table 7.1. The structure of recombinational repair pathways and their back-ups in* E. coli**

	DNA Lesion	
	Daughter-strand gap	**Double-strand end**
Main repair pathway	RecA-RecFOR (recombinational)	RecA-RecBCD (recombinational)
Back-up	RecA-UmuDC (translesion DNA synthesis)	RecBCD (degradation)

local levels. The mechanisms of the global inhibition are better understood in yeast, where the cell-cycle response to DNA damage is regulated by the RAD9/RAD24/RAD53 pathway.[1-3] Yeast cells mutant in these genes proceed with the cell cycle when their DNA is damaged and consequently die. The inhibition of cell cycle by DNA damage allocates sufficient time for repair, while the inhibition of DNA replication on damaged templates prevents massive conversion of one-strand damage into two-strand damage. A similar pathway apparently operates in mammalian cells (discussed in ref. 4), however, the molecular details are not yet available.

Local mechanisms of delaying DNA replication of damaged DNA are better understood for mammalian cells; some of these mechanisms are not found in yeast, but analogous responses could be expected. Single-strand interruptions in DNA attract an abundant enzyme called poly(ADP-ribose)polymerase (PARP), which: (1) binds these interruptions; (2) poly-ADP-ribosylates surrounding proteins and (3) poly-ADP-ribosylates itself and dissociates from DNA. Poly-ADP-ribosylation inhibits the affected proteins, including various components of DNA replication machinery. Possible functions of this short-lived modification are still discussed.[5,6] One is tempted to speculate that PARP-modification keeps active DNA replication away from regions undergoing one-strand repair, especially base-excision repair[4] serving as a transient barricade.

The other damage-responsive element in eukaryotic cells is an abundant protein Ku, which specifically binds to double-strand ends. The bound Ku induces DNA-dependent protein kinase (DNA-PK) that phosphorylates surrounding, mostly DNA-bound, proteins (reviewed in refs. 7-9). The exact role of this protein modification is not known, although cells mutant in Ku subunits are defective in double-strand break repair, suggesting that repair proteins could be activated, while other enzymes that might otherwise aggravate the damage are inhibited. Among the substrates of DNA-PK is p53 tumor suppressor protein, which controls global response to DNA damage, delaying S phase (reviewed in refs. 10, 11). Interestingly, in contrast to the expectation that a mutant in a DNA damage checkpoint should be sensitive to DNA damage, survival of the cells deleted for p53 after DNA damaging treatments is actually better than that of WT cells.[12] The unexpected effect of p53 inactivation suggests that the main task of the p53-pathway is not to help with repair of DNA damage, but to induce the cell-suicide mechanisms (apoptosis) when repair of the damage is likely to change or reverse the cell's developmental program.[11] Apoptosis, which includes fragmentation and degradation of chromosomal DNA, is the last protective measure of multicellular organisms against uncontrollable cell proliferation. Elimination of abnormal cells in multicellular organisms looks somewhat analogous to the RecBCD-promoted degradation of the abnormally-replicating nucleoids in *E. coli* (5.5).

Thus, responses of eukaryotic cells to DNA damage not only increase cells' capacity for repair and damage tolerance, but also inhibit DNA synthesis locally as well as globally. The inhibition of DNA synthesis is important for eukaryotes because multiple replicons in their chromosomes elevate the chances of replication fork encounter with a DNA lesion. The sheer size of a eukaryotic genome makes it difficult to prevent all such encounters though, and the substrates for recombinational repair are always generated. Recombinational repair is detected in eukaryotic cells as diverse as yeast and mammalian somatic cells.

7.1. ORGANIZATION OF RECOMBINATIONAL REPAIR IN YEAST, *SACCHAROMYCES CEREVISIAE*

To say simply that recombinational repair "can be detected" in yeast is to unjustly belittle this workhorse of recombinational repair studies. Yeast offers a happy combination of "the awesome power of yeast genetics" with the most convenient in vivo system for studying recombinational repair reactions "in real time". Besides, meiosis in general and yeast meiosis in particular is a natural "recombinational symphony" due to the requirements of crossing-over for proper chromosome segregation. It is not surprising then that the concept of recombinational repair, having originated in bacterial studies, was first enthusiastically accepted in yeast research.

7.1.1. Yeast RecA-Like Proteins

The first indication of the presence of recombinational repair in a particular organism is the finding in this organism of RecA-like proteins. Budding yeast have four proteins with homology to bacterial RecA proteins. Rad51 protein is 30% homologous to the *E. coli* RecA in its catalytically important central region.[13] In vitro, Rad51 protein assembles around dsDNA in a spiral filament resembling RecA filament[14] and promotes the basic strand exchange reaction, pairing of circular single-stranded DNA with homologous linear duplex DNA.[15] The phenotype of Rad51 mutants correspond to that of would-be mutants in RecA-like protein: they are sensitive to ionizing radiation and defective in interhomolog recombination (reviewed in refs. 16,17). Rad51 expression is induced in response to DNA damage.[13,18] Although it is yet to be shown whether this protein can promote a biologically-relevant strand-invasion reaction, Rad51 is considered to be the yeast synaptase.

When yeast cells undergo meiosis, their DNA suffers site-specific double-strand breaks,[19-21] which are transient in WT cells due to their repair off an intact homolog.[22,23] The recombinational repair of these breaks is thought to ensure intimate interactions between homologous chromosomes which is important for proper chromosomal distribution during the first division of meiosis. The expression of yeast Rad51 protein is elevated during meiosis.[13] In situ hybridization of the yeast cells in meiotic prophase I detects Rad51 protein in several dozens foci along the interacting chromosomes.[24] *rad51* mutants do not form viable spores, which apparently is the consequence of their defect in repair of meiosis-specific double-strand breaks.[13]

S. cerevisiae has at least three other proteins with significant homology to the *E. coli* RecA and to RAD51 proteins (reviewed in ref. 25). Rad55 and Rad57 are required for resistance to ionizing radiation in mitotic cells (reviewed in ref. 16). The *rad55* and *rad57* deficiencies are partially suppressed by overproducing Rad51 protein.[26,27] Expression of the third RecA homolog, DmcI, is restricted to meiosis where this protein is required for repair of meiosis-specific double-strand breaks[28] and can be found in nuclear foci together with RAD51.[24]

Why yeast cells may need several flavors of RecA-like proteins is unclear. Rad55 and Rad57 might, for example, contribute to stability of Rad51 filament in certain conditions. On the other hand, the existence of organisms with several different RecA-like proteins is expected on the basis of the proposed mechanism for homology search by a synaptase filament (2.2.3). It is suggested that a synaptase filament comprises two types of monomers: tight-binders and loose-binders. Tight-binders form the filament backbone, while loose-binders bring other DNAs to align with the inner ssDNA in the filament. In bacteria, the division of labor is virtual rather than structural — a particular RecA monomer can be either a tight-binder or a loose-binder. Perhaps, in eukaryotes the division of labor is structural. For example, Rad51 protein could be a tight-binder forming a backbone of the synaptase filament around inner ssDNA, while Rad55 and Rad57 could be loose-binders responsible for bringing duplex DNA to the filament. In the absence of Rad55 and/or Rad57, high concentrations of Rad51 would allow it to play both roles, similar to bacterial RecA.

7.1.2. Double-Strand End Repair in Yeast

DNA damage repair in yeast is controlled by genes belonging to three epistasis groups (reviewed in ref. 16). The *RAD3* group comprises the genes of excision repair (resistance to UV), the *RAD6* group is responsible for daughter-strand gap repair (resistance to both UV and ionizing radiation) (7.1.3), while the *RAD52* group provides double-strand end repair (resistance to ionizing radiation). Genetic recombination is decreased only in mutants of the genes of the *RAD52* epistasis group (reviewed in ref. 17), indicating that recombinational repair in yeast is used to mend mostly double-strand breaks and disintegrated replication forks.

Double-strand end repair in yeast is characterized both genetically and by physical monitoring of the repair reactions in vivo (reviewed in ref. 29). The presynaptic phase of DSE repair in yeast begins with resection of a double-strand end generating a 3'-overhang (Fig. 7.1). This exonucleolytic degradation is slower in *rad50* mutants, suggesting that Rad50 protein controls it.[30] Two more mutants, *xrs2* and *mre11*, show general phenotypes very similar to the phenotype of *rad50* mutants,[31,32] moreover, Mre11 protein directly interacts with both Rad50 and Xrs2 in vivo.[33] Rad50, Xrs2 and Mre11 might, for example, constitute the exonuclease itself or they might be a part of the yeast damage-responsive element, bound to a double-strand end. The generated single-strand overhang is likely to be immediately complexed by SSB protein (called RPA in yeast); cells with a mutant RPA are sensitive to ionizing radiation.[34] This defect due to a mutant RPA is suppressed by overproduction of RAD52 protein,[34] suggesting that RAD52 may also bind to the single-strand overhang. Indeed, the end resection in *rad52* mutants is excessive and spreads to the single-strand overhang itself,[30] which is consistent with RAD52 binding to the 3'-overhang or mediating other protein(s) binding to it. The binding of RPA and possibly Rad52 to the single-strand overhang is likely to be a prerequisite for synaptase filament formation.

Rad51 catalyses the basic strand transfer reaction in vitro (7.1.1) and is also known to interact with Rad52 both in vitro and in vivo.[13,35] The direct interaction of Rad52 with Rad51 suggests that the assembly of synaptase filament on the single-stranded DNA could be promoted by Rad52 protein (Fig. 7.1). The formation of active synaptase filament in vivo might also require Rad55 and Rad57 RecA-like proteins since they are indispensable for the resistance to ionizing radiation and recombination proficiency. In vivo, Rad55 interacts with both Rad51 and Rad57; the latter two proteins do not interact with each other.[26,27] The only protein left outside this scheme is Rad54 which is, nevertheless, required for reassembly of the chromosomal DNA fragmented by ionizing radiation.[36] Expression of Rad54 is induced by DNA damage;[37] the sequence of Rad54 protein reveals its similarity to DNA-helicases.[38] Rad54 might participate in DSB repair either in the presynaptic phase (as a DNA helicase) or in the postsynaptic phase (for example, helping to restart DNA synthesis at re-established replication forks, like PriA in *E. coli* (5.4.1)).

An interesting difference between recombinational repair in yeast and in *E. coli* is in the phenotypes of null mutants in synaptase genes. While *recA* mutation prohibits almost all homologous recombination in the *E. coli* chromosome, *rad51*, *rad55* or *rad57* mutations have a rather moderate effect on the chromosomal recombination in yeast.[39,40] The yeast mutants that show the levels of recombination deficiency similar to bacterial *recA* mutants are in *rad52* gene,[39] whose product

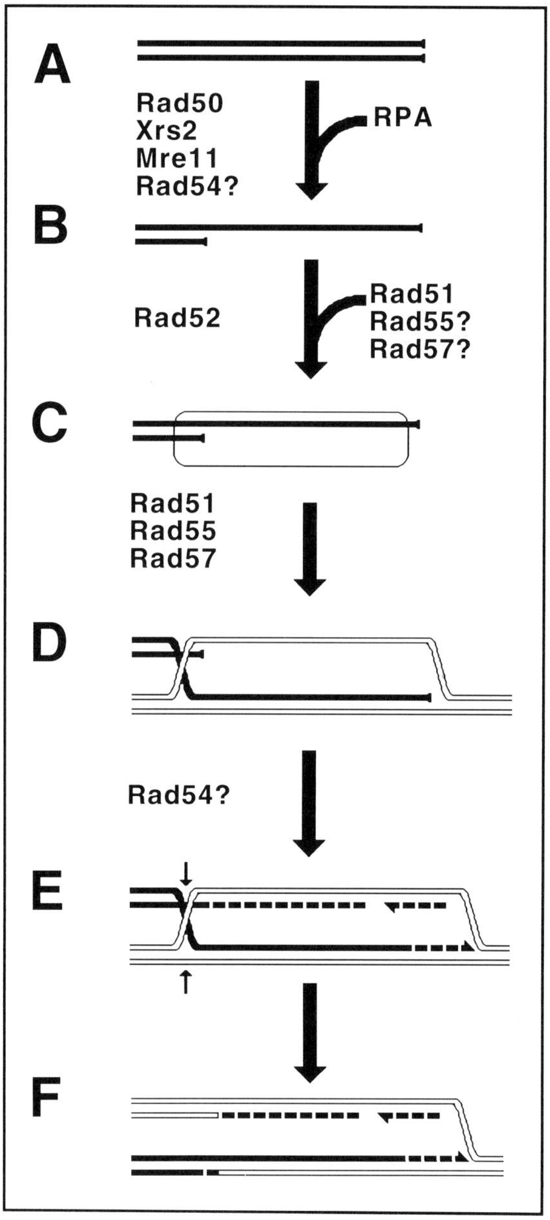

Fig. 7.1. A scheme for double-strand end repair in yeast. Rectangle with rounded corners, synaptase filament, newly-synthesized DNA is shown as interrupted lines, resolution of the Holliday junction is indicated by a pair of small vertical arrows. (A) A blunt double-strand end attracts damage-responsive proteins (Rad50? Xrs2? Mre11?), which arrange its processing (Rad54 participates?). (B) The end is resected to have a 3'-ending single-strand overhang, which is complexed by RPA (single-strand DNA-binding protein). Rad52 is recruited (by the damage-responsive element?) to the RPA-complexed single-strand overhang. (C) Rad52 assists in Rad51 filament assembly on the single-strand overhang. Displacement of Rad52 from the Rad51 filament by Rad55 + Rad57 activates the filament for homology search. (D) The homologous sequence is found and paired with the double-strand end, creating a Holliday junction. (E) The invasion structure is converted (with the help of Rad54?) into a replication fork. (F) The Holliday junction is resolved by yet to be characterized enzymes (only one out of two possibilities is shown), completing the repair reaction.

has no homology to RecA protein. RAD52 protein may promote RAD51-filament assembly on single-stranded DNA complexed by RPA, which is analogous to the assistance of RecF, RecO and RecR proteins in the RecA polymerization on ssDNA complexed with SSB in *E. coli* (4.2.3 and 5.6). In the absence of one of the synaptase proteins, RAD52 might be able to promote formation of partly functional synaptase filaments composed of the remaining two proteins, accounting for the residual recombinational repair.

The model of double-strand end repair in yeast looks similar to the alternative DSE repair mechanisms in *E. coli* (5.6), though there are inevitable differences. For example, RAD51 filament was shown to promote strand exchange in the 3'-to-5' direction, which is opposite to the direction of RecA-promoted strand exchange.[41] Also, double-strand ends in eukaryotic cells are likely to be bound by damage-responsive activities, which might orchestrate the whole repair reaction (at least, promote synaptase filament formation) via modification of the appropriate proteins by the associated protein kinases (7.0). In contrast, recombinational repair in *E. coli* is proposed to be controlled directly by inhibition of DNA synthesis (3.2). The completion of damage repair in eukaryotic cells is monitored by checkpoint mechanisms which inhibit further cell cycle progress until all the lesions are repaired.[42] This makes unnecessary the degradational back-up proposed to play an important role in preventing over-replication in bacterial chromosomes (5.5). Indeed, eukaryotic cells do not have an exonuclease that would match the murderous efficiency of the bacterial ExoV activity (5.3).

The absence of a strong degradative exonuclease in eukaryotic cells makes possible single-strand annealing repair of double-strand breaks between direct repeats in their chromosomes (6.0). In yeast, such repair is independent of Rad51 protein and is dependent on Rad52, Rad1 and Rad10 proteins (reviewed in ref. 29) (Fig. 7.2). The dependence of SSA repair on Rad52 is relieved in long arrays of repeats, again suggesting that Rad52 regulates resection of double-strand ends.[43] RAD1 combines with Rad10 to form an endonuclease which is proposed to clip off unpaired single strand "whiskers".[44] Although SSA repair of double-strand breaks is efficient in yeast, it is unlikely to be of prime importance for this organism due to the lack of repeated sequences in the yeast genome.

7.1.3. Daughter-Strand Gap Repair in Yeast

After UV irradiation, the newly-synthesized DNA in yeast contains gaps indicating that, as in *E. coli*, yeast replisomes are stalled at UV-lesions and reinitiate downstream.[45,46] However, repair of daughter-strand gaps in yeast is predominantly a nonrecombinational process,[46] as also indicated by almost wild type UV-resistance of *rad51*, *rad55* and *rad57* mutants (reviewed in ref. 16). Indeed, daughter-strand

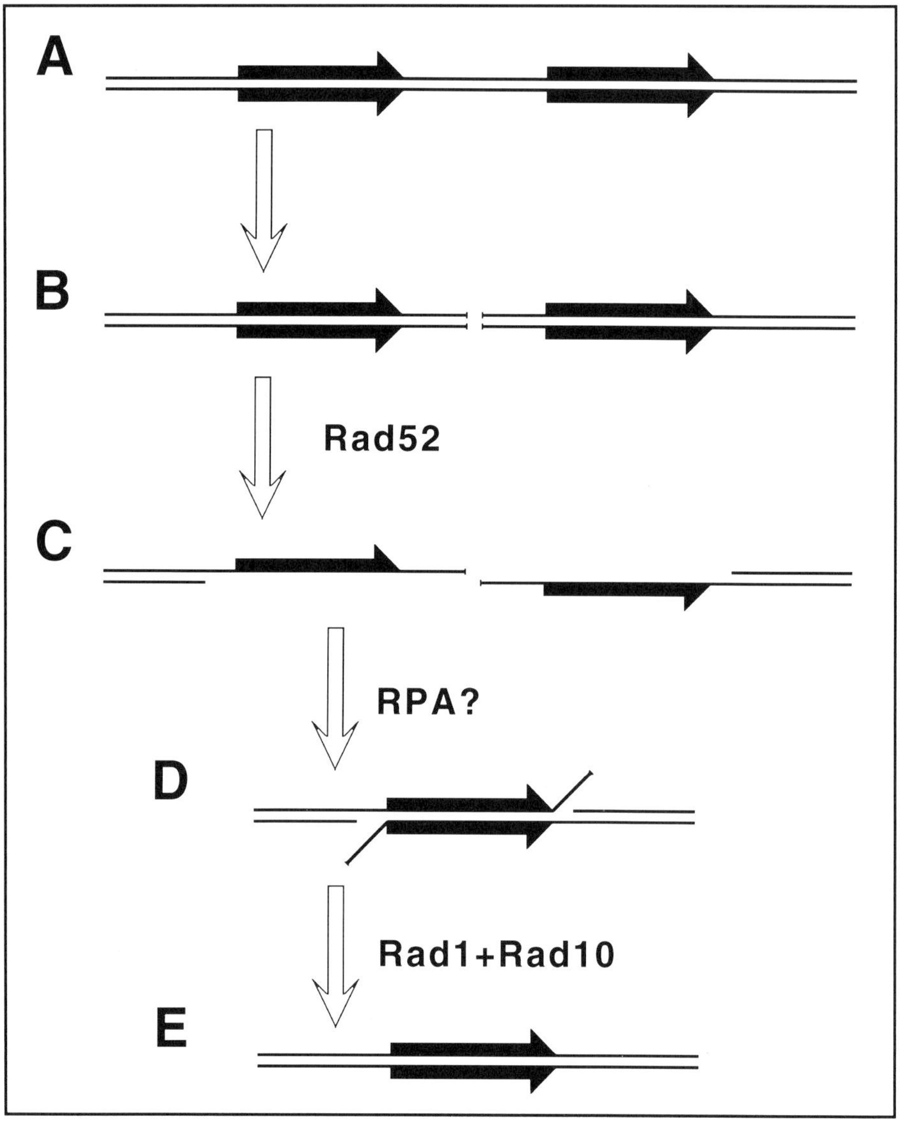

Fig. 7.2. A scheme for single-strand annealing repair in yeast. Direct repeats are shown as black arrows. (A) A chromosome with directly-repeated sequences. (B) A double-strand break occurs between the repeats. (C) The double-strand ends are resected to generate single-strand 3'-overhangs; the resection is controlled by Rad52. (D) The unraveled complementary strands anneal together, reforming repeat-containing duplex; this step could be promoted by RPA, which is known to catalyze annealing of complementary single-strand DNA (reviewed in ref. 81). (E) Rad1/Rad10 endonuclease clips off the single-strand "whiskers", allowing filling-in the single-strand gaps by repair DNA polymerase and sealing nicks by DNA ligase.

gaps in yeast are filled in directly by translesion DNA synthesis controlled by genes of the *RAD18* subgroup of the *RAD6* epistasis group (reviewed in ref. 47). In contrast to a rather inefficient (16%) translesion DNA synthesis past the major UV-lesion in *E. coli* which in addition makes almost 8% mistakes, the same process in yeast is efficient (80%) and accurate, allowing only 0.4% mistakes.[48] *rad6* and *rad18* mutants are very sensitive to UV light, but their UV-resistance is returned back to normal by additional *srs2* mutation,[49] inactivating a DNA helicase.[50] Interestingly, the resulting proficiency in daughter-strand gap repair is now dependent on genes of the Rad52 epistasis group, that is, on recombinational repair.[51] Thus, recombinational repair of daughter-strand gaps in yeast is available, but is normally suppressed by Srs2 helicase. Srs2 helicase apparently interferes with the Rad52-promoted reaction,[52,53] probably preventing assembly of synaptase filaments at daughter-strand gaps. Perhaps, faithful translesion DNA synthesis is cheaper than recombinational repair. In fact, uncontrolled recombinational repair in yeast is evidently poisonous, as suggested by the low viability of *srs2 rad50* double mutants and inviability of *srs2 rad54* double mutants,[54] the latter being suppressed by mutations in *RAD51*, *RAD52*, *RAD55* and *RAD57*.[53] However, if a DNA lesion cannot be bypassed by a modified replisome, recombinational repair is probably activated. Therefore, in contrast to *E. coli*, translesion DNA synthesis is the main route for daughter-strand gap repair in yeast, while recombinational repair is a back-up (Table 7.2).

7.2. RECOMBINATIONAL REPAIR IN HIGHER EUKARYOTES

Besides budding yeast, recombinational repair genes are found in other eukaryotes, including fission yeast, filamentous fungi, insects, amphibians, birds and mammals (reviewed in ref. 25). However, while lower eukaryotes seem to express these proteins continuously, the majority of cell types in multicellular organisms show little expression of recombinational repair proteins. Daughter-strand gaps in DNA of higher

***Table 7.2. The structure of recombinational repair pathways and their back-ups in budding yeast,* S. cerevisiae**

	DNA Lesion	
	Daughter-strand gap	**Double-strand end**
Main repair pathway	Rad6-Rad18 (translesion DNA synthesis)	Rad52-Rad51 (recombinational)
Back-up	Rad52-Rad51 (recombinational)	Rad52-Rad1/Rad10 (SSA repair of DSB)

eukaryotes are repaired mostly by translesion DNA synthesis, whereas repair of double-strand breaks relies on DNA-PK signaling (reviewed in ref. 4). DNA with double-strand breaks transfected into mammalian cells is efficiently repaired by simple ligation or is randomly inserted into the genome (reviewed in ref. 55). If direct repeats are provided on both sides of double-strand ends, cells of higher eukaryotes are able to recombine the transfected linear DNA by single-strand annealing.[56] Perhaps, the role of recombinational repair in cells of higher eukaryotes is mostly to mend disintegrated replication forks, while its scope is restricted to newly-replicated sister duplexes. These limitations should hamper genetic studies of recombinational repair in higher eukaryotes; to facilitate them, sister-chromatid exchange detection might turn out to be useful.

7.2.1. Recombinational Repair Proteins in Higher Eukaryotes

Rad51 homologs are found in fission yeast,[57] fruit fly,[58] frog,[59] chicken,[60] mouse,[57] human[57] and lily,[61] proving to be ubiquitous. Human Rad51 protein is only 27% divergent from yeast Rad51, whereas human, mouse and chicken proteins are almost identical.[62] Interestingly, high expression of *RAD51* genes in frog, chicken and mouse is detected only in reproductive tissues and in the organs of the immune system.[57,59,60] In reproductive tissues (ovaries, testicles) cells undergo meiosis, while cells of the immune system rearrange their antibody or receptor genes. In human, as well as in lily cells undergoing meiosis, Rad51 protein is distributed among several dozens foci along the interacting chromosomes,[61,63] which is analogous to the situation in budding yeast (7.1.1). Similar Rad51-containing foci are observed in a fraction of human somatic cells after DNA-damaging treatments.[63] A homolog of the yeast meiosis-specific RecA-like protein Dmc1, called Lim15, is found in lily;[64] as its yeast counterpart (7.1.1), it is colocalized with lily Rad51 protein during meiosis.[61]

Homologs of budding yeast Rad52 proteins are found in fission yeast,[65] chicken,[66] mouse and human.[67] Like their Rad51 counterparts, chicken, mouse and human Rad52 homologs are highly expressed in reproductive tissues and in the organs of the immune system.[66,67] The presence and expression of Rad51 and Rad52 homologs in higher eukaryotes indicates that the double-strand end pathway of recombinational repair is used throughout eukaryotic kingdom.

7.2.2. Sister Chromatid Exchange as a Possible Indicator of Recombinational Repair in Higher Eukaryotes

Low level of recombination between homologous chromosomes in mitotically dividing cells of higher eukaryotes hampers genetic studies of recombinational repair in cultured cells. In mitotically dividing yeast cells, sister chromatids are preferred over homologs for recombinational repair;[68] this preference is likely to be even higher in mammalian cells due to the greater size of their chromosomes. Crossing-over between sister chromatids cannot be detected genetically unless the chromosomes carry directly repeated segments of DNA. Fortunately, there is a cytological method so sensitive that it allows detection of a single exchange between sister chromatids in big chromosomes of vertebrates. The method is based on differential labeling of sister chromatids (Fig. 7.3); it was successfully used in characterizing sister-chromatid exchange (SCE) in vivo and in a variety of cultured cells in vitro.[69,70]

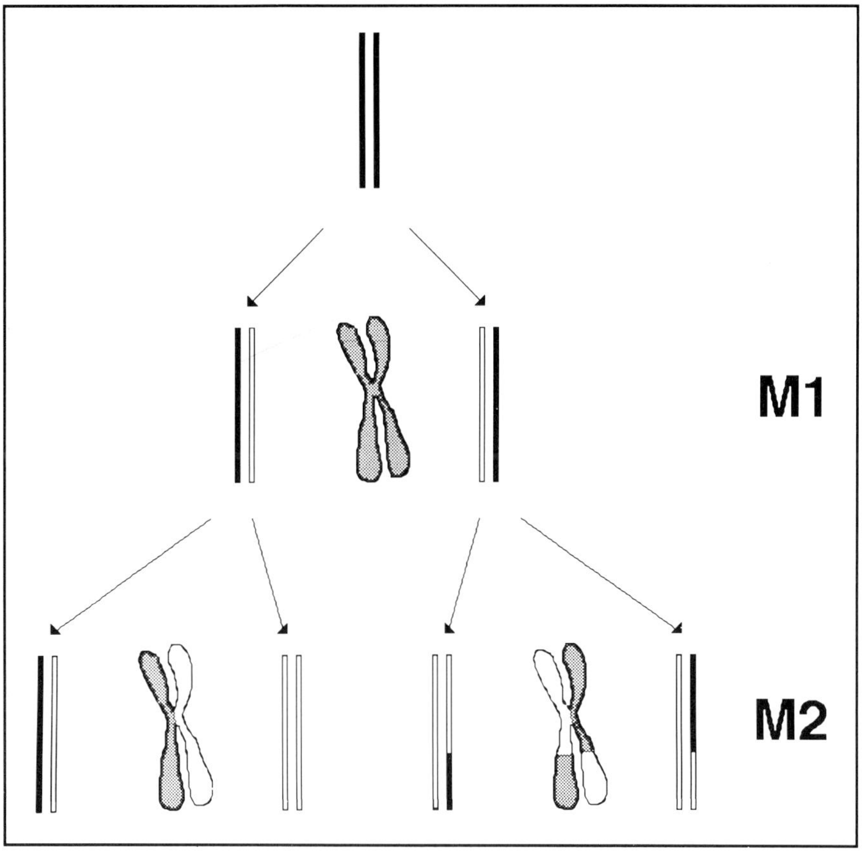

Fig. 7.3. Cytological detection of sister-chromatid exchange. DNA duplexes (single chromatids) are represented by pairs of vertical lines. Black lines designate labeled DNA strands, white lines designate unlabeled strands. M1, metaphase of the first mitosis; M2, metaphase of the second mitosis. Metaphase chromosomes, composed of the two sister chromatids, are shown between the corresponding chromatids with staining pattern reflecting their strand composition. Top — a fully labeled chromatid, duplication of which generates two half-labeled chromatids (center). The resulting metaphase bivalent is uniformly labeled. Bottom left — a subsequent duplication of a half-labeled chromatid generates a half-labeled chromatid and an unlabeled chromatid, which are combined in a half-stained metaphase bivalent. Bottom right — a similar duplication of a half-labeled chromatid accompanied by sister-chromatid exchange produces a characteristic switched staining pattern in the resulting bivalent.

The influence of various DNA-damaging agents as well as some human hereditary conditions on sister-chromatid exchange is thoroughly documented.[69,70] The accumulated evidence unequivocally shows that sister-chromatid exchange is a consequence of DNA replication on a damaged template. In bacteria and in yeast, the two-strand lesions resulting from DNA replication on damaged templates are mended by recombinational repair. However, the lack of recombinational repair mutants in higher eukaryotes and the independence of SCE formation of any other known repair system led the development of models for sister-chromatid exchange in a different direction. Current models invariably picture SCE as a simple break-exchange-join reaction putting emphasis on fortuitous or damage-stimulated coinciding (if not simultaneous) breaks in both sister chromatids at a replication fork with subsequent swapping and reattachment.[71-75] Since the major class of enzymes known to catalyze sequence-nonspecific break-exchange-join reactions are DNA topoisomerases, this development logically culminated in the proposal that SCE is catalyzed by topoisomerases.[75] We noticed that sister-chromatid exchange could be also accounted for by the current models of recombinational repair.

In fact, a number of well-established observations suggest a recombinational repair origin of SCE. The frequency of SCE is generally stimulated by DNA damage. Cross-linking agents are the most efficient stimulators, followed by alkylating agents, UV light and ionizing radiation. As discussed above (7.0), poly(ADP-ribose)polymerase is a ubiquitous enzyme that binds to breaks and ADP-ribosylates a variety of proteins, inhibiting their activities. Sister-chromatid exchange is stimulated by inhibition of poly(ADP-ribose)polymerase (reviewed in ref. 5), suggesting that undetected DNA strand breaks could cause SCE. Indeed, in human cells from patients with Bloom's syndrome, which have abnormal DNA ligase activity, there is a several-fold increase in basal frequency of SCE as well as stronger induction of SCE by DNA-damaging agents (reviewed in refs. 4,70). In human cells deficient in excision repair (from patients with Xeroderma Pigmentosum), the basal frequency of SCE is not changed, but UV light stimulation of SCE formation is greater than in excision repair-proficient cells (reviewed in refs. 4,70), suggesting that unremoved pyrimidine dimers could also cause SCE.

SCE happens during or soon after DNA replication,[76-79] indicating that it is caused by replication fork encounters with unrepaired lesions. The resulting two-strand lesions are either daughter-strand gaps (if the one-strand lesions were of noncoding type) (1.2.1 and 4.1) or disintegrated replication forks (if the one-strand lesions were single-strand interruptions or cross-links) (1.2.1 and 5.1). In bacteria, these two-strand lesions are mended by two major pathways of recombinational repair (chapters 4 and 5). The current models of this repair should be suitable for sister-chromatid exchange as well (Fig. 7.4).

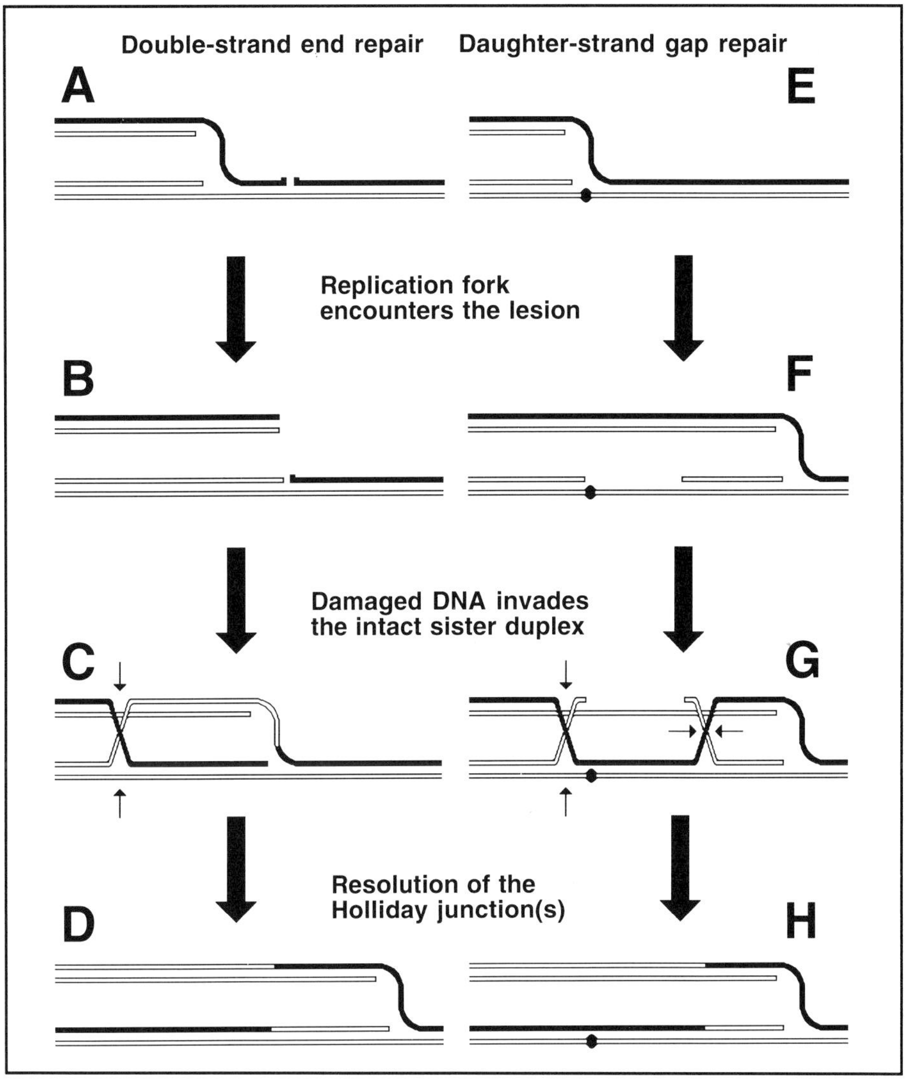

Fig. 7.4. Sister-chromatid exchange could be a consequence of recombinational repair. Recombinational repair is used to mend double-strand ends and daughter-strand gaps. On the left, a scheme describing formation and repair of a double-strand end is featured. (A) Replication of a half-labeled chromatid (central part in Fig. 7.3) containing a single-strand interruption. (B) When a replication fork encounters the interruption, it disintegrates generating a double-strand end[74] (1.1.4). (C) The single-strand interruption is repaired, while the double-strand end with a help of a synaptase filament invades the intact duplex, re-creating a replication fork framework accompanied by a Holliday junction. Resolution of the Holliday junction in the direction shown by the small arrows ensures the formation of a strand exchange, shown in "D". On the right, a scheme of formation and repair of a daughter-strand gap is shown. (E) Replication of a half-labeled chromatid containing an unrepaired one-strand lesion. (F) The replisome is blocked by the lesion and reinitiates downstream, leaving behind a single-strand gap in one of the daughter duplexes. (G) The daughter-strand gap with a help of a synaptase filament invades the intact duplex, generating a pair of Holliday junctions. (H) The gap is filled and the junctions are resolved in the direction shown in "G", dictating the formation of a strand exchange.

Currently, there is no evidence that would favor the recombinational repair models for SCE formation over the existing simple break-join models. The recombinational repair models are more attractive, though, because they are based on well-understood processes of DNA maintenance common in bacteria and lower eukaryotes. Besides, recombinational repair models not only explain the phenomenology of SCE, but also generate specific testable predictions. One such prediction is a depression of sister-chromatid exchange in *rad51* and *rad52* mutant cells. Another prediction is based on the hyper-recombination phenotype of *E. coli* mutants, exhibiting increased number of single-strand interruptions in their DNA.[80] The list includes mutants in DNA ligase, repair DNA polymerase and DNA precursors metabolism. Analogous mutants in higher eukaryotes could be expected to exhibit increased basal level of SCE. Finally, one may predict that combinations of such hyper-SCE mutations with mutations defective in recombinational repair in mammalian cells are nonviable or develop gross chromosomal rearrangements.

SUMMARY

Eukaryotic cells react to DNA damage with local and global responses designed to inhibit cell-cycle progress and DNA synthesis. These preventive measures decrease the chances of a replication fork encounter with unrepaired lesions. However, due to the multiple replicons in the eukaryotic chromosomes, some conversion of one-strand lesions to two-strand lesions is inevitable. Therefore, recombinational repair warrants its place in the eukaryotic repertoire of DNA repair transactions.

Recombinational repair in yeast is used to mend disintegrated replication forks and double-strand breaks. Daughter-strand gaps are repaired preferentially by translesion DNA synthesis, recombinational repair being available as an option (the only option in certain mutants). Mitotically dividing yeast cells express three RecA-like proteins, all three being required for recombinational repair in vivo.

Homologs of the yeast recombinational repair proteins are found in plants, insects and vertebrates, suggesting evolutionary conservation of the recombinational DSE repair pathway. Genetic studies of recombinational repair in mitotic cells of higher eukaryotes are hampered by the genome complexity, low rates of recombination and very efficient end joining. However, if cytologically-detectable sister-chromatid exchange proves to be the consequence of recombinational repair, genetic analysis of recombinational repair mutants in higher eukaryotes will be facilitated.

REFERENCES

1. Weinert TA and Hartwell LH (1988) The *RAD9* gene controls the cell cycle response to DNA damage in *Saccharomyces cerevisiae*. Science 241:317-322.

2. Allen JB, Zhou Z, Siede W, Friedberg EC and Elledge SJ (1994) The *SAD1/RAD53* protein kinase controls multiple checkpoints and DNA damage-induced transcription in yeast. Genes Dev 8:2416-2428.
3. Siede W, Friedberg AS, Dianova I and Friedberg EC (1994) Characterization of G_1 checkpoint control in the yeast *Saccharomyces cerevisiae* following exposure to DNA-damaging agents. Genetics 138:271-281.
4. Friedberg EC, Walker GC and Siede W (1995) DNA Repair and Mutagenesis. Washington, D.C.: ASM Press, 698 p.
5. Cleaver JE and Morgan WF (1991) Poly(ADP-ribose)polymerase: a perplexing participant in cellular responses to DNA breakage. Mutat Res 257:1-18.
6. Lindahl T, Satoh MS and Dianov G (1995) Enzymes acting at strand interruptions in DNA. Phil Trans R Soc Lond 347:57-62.
7. Anderson CW (1993) DNA damage and the DNA-activated protein kinase. Trends Biochem Sci 18:433-437.
8. Jeggo PA, Taccioli GE and Jackson SP (1995) Menage à trois: double strand break repair, V(D)J recombination and DNA-PK. BioEssays 17:949-957.
9. Weaver DT (1995) What to do at an end: DNA double-strand-break repair. Trends Genet 11:388-392.
10. Oliner JD (1993) Discerning the function of p53 by examining its molecular interactions. BioEssays 15:703-707.
11. Elledge RM and Lee W-H (1995) Life and death by p53. BioEssays 17:923-930.
12. Lee JM and Bernstein A (1993) p53 mutations increase resistance to ionizing radiation. Proc Natl Acad Sci USA 90:5742-5746.
13. Shinohara A, Ogawa H and Ogawa T (1992) Rad51 protein involved in repair and recombination in *S. cerevisiae* is a RecA-like protein. Cell 69:457-470.
14. Ogawa T, Yu X, Shinohara A and Egelman EH (1993) Similarity of the yeast Rad51 filament to the bacterial RecA filament. Science 259: 1896-1899.
15. Sung P (1994) Catalysis of ATP-dependent homologous DNA pairing and strand exchange by yeast Rad51 protein. Science 265:1241-1243.
16. Haynes RH and Kunz BA (1981) DNA repair and mutagenesis in yeast. In: Strathern JN, Jones EW and Broach JR, eds. The Molecular Biology of the Yeast *Saccharomyces*. Cold Spring Harbor, N.Y.: Cold Spring Harbor Laboratory, pp 371-414.
17. Petes TD, Malone RE and Symington LS (1991) Recombination in yeast. In: Broach JR, Pringle JR and Jones EW, eds. The molecular and cellular biology of the yeast *Saccharomyces*. I. Genome dynamics, protein synthesis and energetics. Cold Spring Harbor Laboratory Press, pp 407-521.
18. Basile G, Aker M and Mortimer RK (1992) Nucleotide sequence and transcriptional regulation of the yeast recombinational repair gene *RAD51*. Mol Cell Biol 12:3235-3246.

19. Cao L, Alani E and Kleckner N (1990) A pathway for generation and processing of double-strand breaks during meiotic recombination in *S. cerevisiae.* Cell 61:1089-101.
20. Sun H, Treco D and Szostak JW (1991) Extensive 3'-overhanging, single-stranded DNA associated with the meiosis-specific double-strand breaks at the *ARG4* recombination initiation site. 64:1151-1161.
21. Game JC (1992) Pulsed-field gel analysis of the pattern of DNA double-strand breaks in the *Saccharomyces* genome during meiosis. Dev Genet 13:485-497.
22. Padmore R, Cao L and Kleckner N (1991) Temporal comparison of recombination and synaptonemal complex formation during meiosis in *S. cerevisiae.* Cell 66:1239-1256.
23. Schwacha A and Kleckner N (1994) Identification of joint molecules that form frequently between homologs but rarely between sister chromatids during yeast meiosis. Cell 76:51-63.
24. Bishop DK (1994) RecA homologs Dmcl and Rad51 interact to form multiple nuclear complexes prior to meiotic chromosome synapsis. Cell 79:1081-1092.
25. Heyer W-D (1994) The search for the right partner: homologous pairing and DNA strand exchange proteins in eukaryotes. Experientia 50:223-233.
26. Hays SL, Firmenich AA and Berg P (1995) Complex formation in yeast double-strand break repair: participation of RAD51, RAD52, RAD55 and RAD57 proteins. Proc Natl Acad Sci USA 92:6925-6929.
27. Johnson RD and Symington LS (1995) Functional differences and interactions among the putative RecA homologs Rad51, Rad55 and Rad57. Mol Cell Biol 15:4843-4850.
28. Bishop DK, Park D, Xu L and Kleckner N (1992) Dmcl: A meiosis-specific yeast homolog of *E. coli* RecA required for recombination, synaptonemal complex formation, and cell cycle progression. Cell 69:439-456.
29. Haber JE (1995) In vivo biochemistry: physical monitoring of recombination induced by site-specific endonucleases. BioEssays 17:609-620.
30. Sugawara N and Haber JE (1992) Characterization of double-strand break-induced recombination: homology requirements and single-stranded DNA formation. Mol Cel Biol 12:563-575.
31. Ivanov EL, Sugawara N, White CI, Fabre F and Haber JE (1994) Mutations in *XRS2* and *RAD50* delay but do not prevent mating-type switching in *Saccharomyces cerevisiae.* Mol Cell Biol 14:3414-3425.
32. Johzuka K and Ogawa H (1995) Interaction of Mre11 and Rad50: two proteins required for DNA repair and meiosis-specific double-strand break formation in *Saccharomyces cerevisiae.* Genetics 139:1521-1532.
33. Ogawa H, Johzuka K, Nakagawa T, Leem S-H and Hagihara AH (1995) Functions of the yeast meiotic recombination genes, *MRE11* and *MRE2.* Adv Biophys 31:67-76.
34. Firmenich AA, Elias-Arnanz M and Berg P (1995) A novel allele of *Saccharomyces cerevisiae RFA1* that is deficient in recombination and repair and suppressible by *RAD52.* Mol Cell Biol 15:1620-1631.

35. Milne G and Weaver D (1993) Dominant negative alleles of *RAD52* reveal a DNA repair/recombination complex including Rad51 and Rad52. Genes Dev 7:1755-1765.
36. Budd M and Mortimer RK (1982) Repair of double-strand breaks in a temperature conditional radiation-sensitive mutant of *Saccharomyces cerevisiae.* Mutat Res 103:19-24.
37. Cole GM, Schild D, Lovett ST and Mortimer RK (1987) Regulation of *RAD54-* and *RAD52-lacZ* gene fusions in *Saccharomyces cerevisiae* in response to DNA damage. Mol Cell Biol 7:1078-1084.
38. Emery HS, Schild D, Kellogg DE and Mortimer RK (1991) Sequence of *RAD54*, a *Saccharomyces cerevisiae* gene involved in recombination and repair. Gene 104:103-106.
39. Rattray AJ and Symington LS (1994) Use of a chromosomal inverted repeat to demonstrate that the *RAD51* and *RAD52* genes of *Saccharomyces cerevisiae* have different roles in mitotic recombination. Genetics 138:587-595.
40. Rattray AJ and Symington LS (1995) Multiple pathways for homologous recombination in *Saccharomyces cerevisiae.* Genetics 139:45-56.
41. Sung P and Robberson DL (1995) DNA strand exchange mediated by a Rad51-ssDNA nucleoprotein filament with polarity opposite to that of RecA. Cell 82:453-461.
42. Bennett CB, Lewis AL, Baldwin KK and Resnick MA (1993) Lethality induced by a single site-specific double-strand break in a dispensable yeast plasmid. Proc Natl Acad Sci USA 90:5613-5617.
43. Ozenberger BA and Roeder GS (1991) A unique pathway of double-strand break repair operates in tandemly repeated genes. Mol Cel Biol 11:1222-1231.
44. Bardwell AJ, Bardwell L, Tomkinson AE and Friedberg EC (1994) Specific cleavage of model recombination and repair intermediates by the yeast Rad1-Rad10 DNA endonuclease. Science 265:2082-2085.
45. Prakash L (1981) Characterization of postreplication repair in *Saccharomyces cerevisiae* and effects of *rad6, rad18, rev3* and *rad52* mutations. Mol Gen Genet 184:471-478.
46. Resnick MA, Boyce J and Cox B (1981) Postreplication repair in *Saccharomyces cerevisiae.* J Bacteriol 146:285-290.
47. Lawrence C (1994) The *RAD6* DNA repair pathway in *Saccharomyces cerevisiae*: What does it do, and how does it do it. BioEssays 16:253-258.
48. Gibbs PEM, Kilbey BJ, Banerjee SK and Lawrence CW (1993) The frequency and accuracy of replication past a thymine-thymine cyclobutane dimer are very different in *Saccharomyces cerevisiae* and *Escherichia coli.* J Bacteriol 175:2607-2612.
49. Lawrence CW and Christensen RB (1979) Metabolic suppressors of trimethoprim and ultraviolet sensitivities of *Saccharomyces cerevisiae rad6* mutants. J Bacteriol 139:866-876.
50. Rong L and Klein HL (1993) Purification and characterization of the SRS2 DNA helicase of the yeast *Saccharomyces cerevisiae.* J Biol Chem 268:1252-1259.

51. Schiestl RH, Prakash S and Prakash L (1990) The *SRS2* suppressor of *rad6* mutations of *Saccharomyces cerevisiae* acts by channeling DNA lesions into the *RAD52* DNA repair pathway. Genetics 124:817-831.
52. Milne GT, Ho T and Weaver DT (1995) Modulation of *Saccharomyces cerevisiae* DNA double-strand break repair by SRS2 and RAD51. Genetics 139:1189-1199.
53. Schild D (1995) Suppression of a new allele of the yeast *RAD52* gene by overexpression of *RAD51*, mutations in *SRS2* and *CCR4*, or mating-type heterozygosity. Genetics 140:115-127.
54. Palladino F and Klein HL (1992) Analysis of mitotic and meiotic defects in *Saccharomyces cerevisiae SRS2* DNA helicase mutants. Genetics 132:23-37.
55. Roth D and Wilson J (1988) Illegitimate recombination in mammalian cells. In: Kucherlapati R and Smith GR, eds. Genetic Recombination. Washington, D.C.: American Society for Microbiology, pp 621-653.
56. Lin F-LM, Sperle K and Sternberg N (1990) Intermolecular recombination between DNAs introduced into mouse L cells is mediated by a nonconservative pathway that leads to crossover products. Mol Cell Biol 10:103-112.
57. Shinohara A, Ogawa H, Matsuda Y, Ushio N, Ikeo K and Ogawa T (1993) Cloning of human, mouse and fission yeast recombination genes homologous to *RAD51* and *recA*. Nature Genet 4:239-243.
58. Akaboshi E, Inoue Y and Ryo H (1994) Cloning of the cDNA and genomic DNA that correspond to the *recA*-like gene of *Drosophila melanogaster*. Jpn J Genet 69:663-670.
59. Maeshima K, Morimatsu K, Shinohara A and Horii T (1995) *Rad51* homologues in *Xenopus laevis*: two distinct genes are highly expressed in ovary and testis. Gene 160:195-200.
60. Bezzubova O, Shinohara A, Mueller RG, Ogawa H and Buerstedde J-M (1993) A chicken Rad51 homologue is expressed at high levels in lymphoid and reproductive organs. Nucleic Acids Res 21:1577-1580.
61. Terasawa M, Shinohara A, Hotta Y, Ogawa H and Ogawa T (1995) Localization of RecA-like recombination proteins on chromosomes of the lily at various meiotic stages. Genes Dev 9:925-934.
62. Ogawa T, Shinohara A, Nabetani A, Ikeya T, Yu X, Egelman EH and Ogawa H (1993) RecA-like recombination proteins in eukaryotes: functions and structures of *RAD51* genes. Cold Spring Harbor Symp Quant Biol 58:567-576.
63. Haaf T, Golub EI, Reddy G, Radding CM and Ward DC (1995) Nuclear foci of mammalian Rad51 recombination protein in somatic cells after DNA damage and its localization in synaptonemal complex. Proc Natl Acad Sci USA 92:2298-2302.
64. Kobayashi T, Hotta Y and Tabata S (1993) Isolation and characterization of a yeast gene which is homologous with a meiosis-specific cDNA from a plant. Mol Gen Genet 237:225-232.
65. Ostermann K, Lorentz A and Schmidt H (1993) The fission yeast *rad22* gene, having a function in mating-type switching and repair of DNA dam-

ages, encodes a protein homologous to Rad52 of *Saccharomyces cerevisiae.* Nucleic Acids Res 21:5940-5944.
66. Bezzubova OY, Schmidt H, Ostermann K, Heyer W-D and Buerstedde J-M (1993) Identification of a chicken *RAD52* homologue suggests conservation of the *RAD52* recombination pathway throughout the evolution of higher eukaryotes. Nucleic Acids Res 21:5945-5949.
67. Muris DRF, Bezzubova O, Buerstedde J-M,et al. (1994) Cloning of human and mouse genes homologous to *RAD52*, a yeast gene involved in DNA repair and recombination. Mutat Res 315:295-305.
68. Kadyk LC and Hartwell LH (1992) Sister chromatids are preferred over homologs as substrates for recombinational repair in *Saccharomyces cerevisiae.* Genetics 132:387-402.
69. Kato H (1977) Spontaneous and induced sister chromatid exchanges as revealed by the BUdR-labeling method. Int Rev Cytol 49:55-97.
70. Latt SA (1981) Sister chromatid exchange formation. Ann Rev Genet 15:11-55.
71. Shafer DA (1977) Replication bypass model of sister chromatid exchanges and implications for Bloom's syndrome and Fanconi's anemia. Hum Genet 39:177-190.
72. Ishii Y and Bender MA (1980) Effects of inhibitors of DNA synthesis on spontaneous and ultraviolet light-induced sister-chromatid exchanges in chinese hamster cells. Mutat Res 79:19-32.
73. Painter RB (1980) A replication model for sister-chromatid exchange. Mutat Res 70:337-341.
74. Saffhill R and Ockey CH (1985) Strand breaks arising from the repair of the 5-bromodeoxyuridine-substituted template and methyl methanesulfonate-induced lesions can explain the formation of sister-chromatid exchanges. Chromosoma 92:218-224.
75. Dillehay LE, Jacobson-Kram D and Williams JR (1989) DNA topoisomerases and models of sister-chromatid exchange. Mutat Res 215:15-23.
76. Kato H (1974) Possible role of DNA synthesis in formation of sister chromatid exchange. Nature 252:739-741.
77. Wolff S, Bodycote J and Painter RB (1974) Sister chromatid exchanges induced in chinese hamster cells by UV irradiation of different stages of the cell cycle: the necessity for cells to pass through S. Mutat Res 25:73-81.
78. Kato H (1977) Mechanisms for sister chromatid exchanges and their relation to the production of chromosomal aberrations. Chromosoma 59:179-191.
79. Latt SA and Loveday KS (1978) Characterization of sister chromatid exchange induction by 8-methoxypsoralen plus near UV light. Cytogenet Cell Genet 21:184-200.
80. Kuzminov A (1995) Collapse and repair of replication forks in *Escherichia coli.* Mol Microbiol 16:373-384.
81. Kowalczykowski SC and Eggleston AK (1994) Homologous pairing and DNA strand-exchange proteins. Annu Rev Biochem 63:991-1043.

CONCLUSION

As noted in the preface of the book edited by K.B. Low (see Selected reading), metabolism of DNA rests on the three "R's". Replication multiplies DNA, one-strand Repair counteracts unwanted DNA modification, while Recombination associates previously separated DNA sequences generating new combinations. Where is the place for Recombinational Repair in this triangle? The substrates for recombinational repair arise mostly when DNA replication encounters unrepaired lesions in template DNA; such aggravated lesions are dealt with by the machinery of homologous recombination, with the goal to make the lesions subject to one-strand repair again. We see that the strategy of recombinational repair is to restore the original pairing in the damaged DNA region disrupted by DNA replication; in essence, it is DNA replication in reverse. However, strand invasion not only undoes what replication has done, but also creates new possibilities for DNA replication. Thus, on the one hand, there is a tight relationship between replication and recombination, while on the other hand, there is a relay relationship between recombinational repair and one-strand repair. Therefore, Recombinational Repair assumes a central position in DNA metabolism, unifying Replication, one-strand Repair and Recombination in a single system of metabolic pathways.

SELECTED READING

These rich sources of experimental facts and theoretical concepts were frequently used by the author but rarely cited, so they deserve an additional mentioning. They can be also recommended for alternative views on recombination and repair.

Friedberg, E.C., Walker, G.C. and Siede, W. (1995) DNA Repair and Mutagenesis. ASM Press, Washington, D.C., 698 p.

An up-to-date treatise on one-strand DNA repair in both prokaryotes and eukaryotes. Also contains updated information on recombinational repair and cellular responses to DNA damage.

Kucherlapati, R. and Smith, G.R., eds (1988) Genetic Recombination. American Society for Microbiology, Washington, D.C., 731 p.

Low, K.B., ed (1988) The Recombination of Genetic Material. Academic Press, Inc., 506 p.

These two older but precious collections of individual chapters on various aspects of recombination complement each other and will be useful for years to come.

Petes, T.D., Malone, R.E. and Symington, L.S. (1991) Recombination in yeast. In: The Molecular and Cellular Biology of the Yeast *Saccharomyces.* volume 1. Genome Dynamics, Protein Synthesis and Energetics. Broach, J.R., Pringle, J.R. and Jones, E.W. eds, Cold Spring Harbor Laboratory Press, 407-521.

The biggest among recent reviews on homologous recombination in yeast.

Kowalczykowski, S.C., Dixon, D.A., Eggleston, A.K., Lauder, S.D. and Rehrauer, W.M. (1994) Biochemistry of Homologous Recombination in *Escherichia coli.* Microbiological Reviews 58: 401-465.

A treatise on genes and enzymes of homologous recombination in E. coli.

INDEX

Page numbers in italics denote figures (f) or tables (t).

Neuroscience Intelligence Unit

Available and Upcoming Titles

- ❒ Neurodegenerative Diseases and Mitochondrial Metabolism
  M. Flint Beal, Harvard University
- ❒ Molecular and Cellular Mechanisms of Neostriatum
  Marjorie A. Ariano and D. James Surmeier, Chicago Medical School
- ❒ Ca^{2+} Regulation By Ca^{2+}-Binding Proteins in Neurodegenerative Disorders
  Claus W. Heizmann and Katharina Braun, University of Zurich, Federal Institute for Neurobiology, Magdeburg
- ❒ Measuring Movement and Locomotion: From Invertebrates to Humans
  Klaus-Peter Ossenkopp, Martin Kavaliers and Paul Sanberg, University of Western Ontario and University of South Florida
- ❒ Triple Repeats in Inherited Neurologic Disease
  Henry Epstein, University of Texas-Houston
- ❒ Cholecystokinin and Anxiety
  Jacques Bradwejn, McGill University
- ❒ Neurofilament Structure and Function
  Gerry Shaw, University of Florida
- ❒ Molecular and Functional Biology of Neurotropic Factors
  Karoly Nikolics, Genentech
- ❒ Prion-related Encephalopathies: Molecular Mechanisms
  Gianluigi Forloni, Istituto di Ricerche Farmacologiche "Mario Negri"-Milan
- ❒ Neurotoxins and Ion Channels
  Alan Harvey, A.J. Anderson and E.G. Rowan, University of Strathclyde
- ❒ Analysis and Modeling of the Mammalian Cortex
  Malcolm P. Young, University of Oxford
- ❒ Free Radical Metabolism and Brain Dysfunction
  Irène Ceballos-Picot, Hôpital Necker-Paris
- ❒ Molecular Mechanisms of the Action of Benzodiazepines
  Adam Doble and Ian L. Martin, Rhône-Poulenc Rorer and University of Alberta
- ❒ Neurodevelopmental Hypothesis of Schizophrenia
  John L. Waddington and Peter Buckley, Royal College of Surgeons-Ireland
- ❒ Synaptic Plasticity in the Retina
  H.J. Wagner, Mustafa Djamgoz and Reto Weiler, University of Tübingen
- ❒ Non-classical Properties of Acetylcholine
  Margaret Appleyard, Royal Free Hospital-London
- ❒ Molecular Mechanisms of Segmental Patterning in the Vertebrate Nervous System
  David G. Wilkinson, National Institute of Medical Research-UK
- ❒ Molecular Character of Memory in the Prefrontal Cortex
  Fraser Wilson, Yale University